Understanding Species

Are species worth saving? Can they be resurrected by technology? What is the use of species in biomedicine? These questions all depend on a clear definition of the concept of 'species', yet biologists have long struggled to define this term.

In this accessible book, John S. Wilkins provides an introduction to the concept of 'species' in biology, philosophy, ethics, policy making and conservation. Using clear language and easy-to-understand examples throughout, the book provides a history of species and why we use them. It encourages readers to appreciate the philosophical depth of the concept as well as its connections to logic and science.

For any interested reader, this short text highlights the complexities of a single idea in biology, the problems with the concept of 'species', and the benefits of it in helping us to answer the bigger questions and understand our living world.

John S. Wilkins teaches at the School of Historical and Philosophical Studies, University of Melbourne, Australia. His main research interests include the philosophy and history of biology and classification, and the cognitive science of religion. He is the author of many books, including *Species: The Evolution of the Idea* (CRC Press, 2018).

The **Understanding Life** series is for anyone wanting an engaging and concise way into a key biological topic. Offering a multidisciplinary perspective, these accessible guides address common misconceptions and misunderstandings in a thoughtful way to help stimulate debate and encourage a more in-depth understanding. Written by leading thinkers in each field, these books are for anyone wanting an expert overview that will enable clearer thinking on each topic.

Series Editor: Kostas Kampourakis http://kampourakis.com

Published titles:

Understanding Evolution	Kostas Kampourakis	9781108746083
Understanding Coronavirus	Raul Rabadan	9781108826716
Understanding Development	Alessandro Minelli	9781108799232
Understanding Evo-Devo	Wallace Arthur	9781108819466
Understanding Genes	Kostas Kampourakis	9781108812825
Understanding DNA Ancestry	Sheldon Krimsky	9781108816038
Understanding Intelligence	Ken Richardson	9781108940368
Understanding Metaphors in the Life Sciences	Andrew S. Reynolds	9781108940498
Understanding Cancer	Robin Hesketh	9781009005999
Understanding How Science Explains the World	Kevin McCain	9781108995504
Understanding Race	Rob DeSalle and Ian Tattersall	9781009055581
Understanding Human Evolution	Ian Tattersall	9781009101998
Understanding Human Metabolism	Keith N. Frayn	9781009108522
Understanding Fertility	Gab Kovacs	9781009054164
Understanding Forensic DNA	Suzanne Bell and John M. Butler	9781009044011
Understanding Natural Selection	Michael Ruse	9781009088329
Understanding Life in the Universe	Wallace Arthur	9781009207324
Understanding Species	John S. Wilkins	9781108987196

Forthcoming:

Understanding Species

JOHN S. WILKINS
University of Melbourne

CAMBRIDGE
UNIVERSITY PRESS

CAMBRIDGE
UNIVERSITY PRESS

Shaftesbury Road, Cambridge CB2 8EA, United Kingdom

One Liberty Plaza, 20th Floor, New York, NY 10006, USA

477 Williamstown Road, Port Melbourne, VIC 3207, Australia

314–321, 3rd Floor, Plot 3, Splendor Forum, Jasola District Centre,
New Delh–110025, India

103 Penang Road, #05–06/07, Visioncrest Commercial, Singapore 238467

Cambridge University Press is part of Cambridge University Press & Assessment,
a department of the University of Cambridge.

We share the University''s mission to contribute to society through the pursuit of
education, learning and research at the highest international levels of excellence.

www.cambridge.org
Information on this title: www.cambridge.org/9781108987196

DOI: 10.1017/9781108982764

First published 2023

A catalogue record for this publication is available from the British Library.

*A Cataloging-in-Publication data record for this book is available from the Library of
Congress.*

ISBN 978-1-108-98719-6 Paperback

'The species problem is a vexing and important one, and John Wilkins has done more than anyone else to dig into its history and integrate it with philosophy past and present. Thus he was the perfect author for this book, which is a wonderful, accessible entryway to the diverse set of issues bearing on why species have been such a "thing" for 2,000 years. My own conclusion is to follow Darwin and acknowledge the species *rank* is a meaningless human construct – the full tree of life is what matters, not just the single level within it arbitrarily called species. But to decide whether to agree with me or not, you need to absorb the content in this book.'

Brent D. Mishler, author of *What, If Anything, Are Species?* (2021),
Distinguished Professor of Integrative Biology,
University of California, Berkeley

'The species problem is one of the most complex issues in evolutionary biology and philosophy of biology, and not many would have succeeded in producing a comprehensive overview of it and doing justice to both science and philosophy. Written by one of the most eminent scholars in the field, *Understanding Species* is an informative and, due to the author's eloquent writing style, at the same time also very entertaining read. It both quenches your thirst for knowledge and makes you want to dive deeper into the topic. What more can you ask of a book? Highly recommended!'

Frank E. Zachos, Natural History Museum Vienna, Austria.
Author of *Species Concepts in Biology* (2016)

'A species is like jazz: you know one when you meet it, but on closer inspection it's very hard to define. In this engaging book, John Wilkins guides us deftly through the philosophical minefield of what species are, how you recognise them, and how trying to find definitions for species is increasingly important for science and conservation.'

Henry Gee, author of *A (Very) Short History of Life on Earth* (2021)

'This book is a stunning achievement, and I think nobody other than Wilkins could have tied together the disparate perspectives needed to write it. Species problems are notoriously thorny and multidisciplinary, yet Wilkins manages to shine great light on them. Most impressively, he does this in ways that many people, rather than just species experts, can understand, engage and enjoy. The writing is snappy, the choice of topics smart, and the rewards for readers will be many.'

Matthew J. Barker, Associate Professor of Philosophy,
Concordia University, Montréal

Contents

Foreword

Everyone knows what a species is, don't they? The bear, the wolf, the shark, etc. So why a whole book on species? Well, the answer is that not everyone really knows what a species is; not because people are ignorant, but because defining species is far from simple and straightforward. Furthermore, and even more counterintuitively for us, there are no exemplars of species, nor sets of distinctive features that we can use to distinguish among them. There is no exemplar of any species. There is no such thing as 'the bear', 'the wolf' or 'the shark', but a variety of bears, wolves and sharks; they share some common features but exhibit an enormous variation in others. As John Wilkins explains in this fascinating book, humans have been preoccupied with classifying organisms around them since ancient times. Yet there has never been a single best way to do this, or to define species, which is the fundamental unit of this classification. Wilkins explains why there exist different definitions that can be in competition with each other or can be consistent with each other. Most importantly, he shows that depending on which species definition we use, we can end up with very different results with respect to classification. Yet, despite problems such as these, species is a useful concept in science. And as Wilkins concludes, understanding is not about acquiring a single true answer or definition. Rather, it is becoming aware of the different uses of a concept, and the different contexts of this use. Reading this book will make you feel that often understanding is about realising how much we do not know. But this is exactly the pleasure of understanding: realising that there is more out there than simplistic accounts can provide. The more you come to know, the more you realise how much else you do not know. Reading this book will make you experience this rewarding feeling.

Kostas Kampourakis, Series Editor

Preface

As a child I asked myself (not my parents or teachers) what a dog or a cat was, or what the pets known as 'carpet snakes' (reticulated pythons) were, and how they differed from other kinds of animals. What I was asking myself at age six, as so many children do at that age, was why cats and tigers are different kinds, or wolves, foxes and dingoes, and so on, and why breeds of dogs sometimes varied so much that one would instinctively group a wolf and an Alsatian together rather than a pug and an Alsatian (I am still not sure if the latter two are in the same group). This is the question of what a species is. Confused, and moving on to space rockets, I promptly forgot about it until I read a paper some 35 years later by one of the founders of modern philosophy of biology, David Hull, and decided to make that my PhD topic.

This is a book for the interested reader. It is not for specialists, nor is it only for those who have studied biology. If you are focused enough, I hope this book, and the readings for each chapter, will give you a good entrée to the smorgasbord of ideas surrounding species. The field is enormous, and many scientists and philosophers feel the need to add to it. This book is a summary of this plethora of options and arguments, and I aim to thread the reader's attention through the issues. I hope it will also help those interested in conservation policy, and in the ideas and the processes of science. Most of all I hope that nobody comes away feeling that at last they have got the idea of species down pat. There is no easy answer, but there are many good questions.

This book, as with all my books, has relied upon the kindnesses of many strangers and friends. First, thanks to Kostas Kampourakis for the opportunity to write this book, and to Jessica Papworth, Olivia Boult and Jenny van der

Meijden at Cambridge University Press, and Lindsay Nightingale, for a really thorough reading and suggestions. I owe many biologists, but especially David Williams at the NHM, Brent Mishler at UC Berkeley, and Frank Zachos at Natural History Museum in Vienna; and many philosopher-historians, especially Matt Barker, Matt Chew and particularly Joachim Dagg, who have been very generous with their reading of drafts, and Jay Odenbaugh for advice. Paul Griffiths gave me my chance to do research in two postdoctoral fellowships, so he gets 10% of the credit and blame. I am also indebted to @Grrlscientist for some suggestions. Thanks especially to my partner Alexis for her tolerance of the writing of this book and of my general lack of attention to anything else for over a year. I'm going to need a new excuse now, though.

A Note

In this book I use *italics* to refer to the words and concepts, and roman text to refer to the things those words are attached to. This includes the formal names of things like species as well as words. We need to do this so that we do not make the common mistake of taking a noun or name as reason to think that there is some real thing that the noun or name refers to. *Unicorn* is a word, a name, but there are no unicorns. Philosophy students know this as the *use–mention* distinction.

1 How Species Matter

There are several 'enigmatic canid' species in North America. One of them is the *red wolf* (*Canis rufus*, Figure 1.1), and another is the *Great Lakes Wolf*. Red wolves are seriously endangered, with a re-released population in North Carolina and breeding programmes being the last populations. Red wolves weren't even studied closely until the 1960s, after having been hunted nearly to extinction in the nineteenth and twentieth centuries.

In 1966, as part of the burgeoning awareness of human impact on wildlife extinctions, the United States passed the *Endangered Species Preservation Act*, later revised in 1973 as the *Endangered Species Act* (ESA) under President Nixon. Under these Acts, the unit of conservation is the species, and any population that does not meet the criteria of specieshood is not worthy of having resources allocated to its conservation.

As a result, much argument has been had about whether the red wolf is a species, a subspecies, or just a population of either coyotes (*Canis latrans*) or the grey wolves (*C. lupus*). Initially, an argument was made by conservation biologists in 1996 that red wolves were 'just' hybrids of these two 'original' species, which meant they did not qualify for these resources. This was rebutted by Ronald Nowak, a Fisheries and Wildlife biologist, who is most responsible for the prominence of the red wolf in conservation debates and policy making.

Hybridisation is, in fact, a common way that the evolution of new species occurs. It can happen in two ways: usually remnant individuals from a population that has been made almost extinct will mate with the nearest approximation. As canids (the 'wolf–dog' group) are widespread and share

Figure 1.1 A typical red wolf.

much of their genetic and developmental machinery, sometimes the hybrids are fertile, and so genetic material is moved from one species to another in a process called *introgression* (the passing of genetic material from one population to another, obviously via mating and having fertile progeny). This means the recipient species now has more genetic variation and can evolve in response to environmental challenges into what is a new species. This evolution into new species becomes evident if a population can no longer interbreed when it gets back in contact with the original species. As we will see, this happened with our ancestors.

The other way is for the hybrids to be different enough from both of the 'parental' species that they form a new species immediately (in geological terms; it could take hundreds or thousands of years). In some plants (e.g., ferns) this is common. It's harder in animals and harder still in vertebrates, but it does happen in lizards and some fishes, as we will see. Or they could just form a subspecies, with a different *phenotype* (the term biologists use for the characteristics that are the products of the organism's *genotype*, which is the entire suite of genetic resources it has) from the parental species but more closely similar to and likely to interbreed with one of the parents.

That wolves interbreed with coyotes is not at issue. The question in the light of the ESA is whether the hybrid form qualifies as its own species. There are three or four major hypotheses: the red wolves are a species derived from a common ancestor with the grey wolf, or they are a hybrid species, or they are a subspecies of one or another of the grey wolves or coyotes, or they are a regional population of one of them. Recent molecular genetic work tends to support the idea that the red wolf is a hybrid. Whether or not this makes it a true species depends upon the species definition one adopts. Depending on which species most of its genetic material comes from, and which species it mates with more successfully, it may be considered a subspecies. Or we may have to revise our notion of species altogether, along with the motivating concepts of conservation biology such as biodiversity and evolutionary uniqueness.

Natural historians (the term for those who studied earthly phenomena in the pre-modern era, including what we now call biologists and geologists, etc.) thought of the world ascending statically and in gradations from inanimate rocks through to nearly angelic humans, with no gaps or separate kinds, and if we haven't found the intermediates locally – the 'missing links' – then it's just that we haven't explored everywhere. In fact, until the eighteenth century, the idea that kinds of animals went extinct was quite literally unimaginable. A beneficent God would not, could not, allow his creations to disappear. This idea was called – in Latin – the *scala naturae*, or in English, the ladder of nature. Historians call it *the Great Chain of Being*. This expected continuity was indeed why many natural historians thought that there were only individuals and not species.

But life *is* characteristically clustered and clumped together at all scales, at least in our world. If there is a 'carpet' of life, it is very wrinkled with many intervening bare patches. No matter what we call them, there are groups of things that live, and there are groups of their parts considered in terms of their structures (both organs and genes). So, if we are to understand the living world about us, and to know what we are dealing with, we need to get a grip on the concepts that we use. Imagining scenarios like the carpet above is a good way to stress-test our intuitions. But there is a better way – we can look at what species are and what they do within science.

The Meaning of *Species*

Species has many traditional meanings and definitions. Chapter 4 will give the short history, and there's a detailed historical and philosophical account in my 2018 book if you want to follow matters further. Still, knowing how ideas and terms were used in the past, while it explains many facets of the use of those words both by specialists and in ordinary use, doesn't help us in how they should be used and for what purposes, any more than the etymology of a dictionary tells us how to use a word today.

However, to consider the purpose of *species* in biology and the wider world, we need to know what they are. And therein lies the rub. Despite school and university textbooks and numerous popular books that treat the idea as uncontroversial, biologists do not agree with one another about what *species* are. And they do not agree in the extreme. By my count, there are around 27 or 28 species 'concepts', up from 22 in a famous 1997 paper by Richard Mayden, an ichthyologist in Saint Louis. Recently, Frank Zachos, a mammalogist in Vienna, has claimed 32 definitions. Since his book, at least three new definitions have been proposed, though they may not really be new. To confuse matters, specialists will also talk about *species concepts* (specifications of the diagnostic characters) of a particular group, say, colobus monkeys or black salamanders. A few words are needed if we aren't to get too tangled up.

First, having a 'concept' does not need to include having a definition. If you disagree with me here (and are not a specialist) try to define your concept of 'dog' in a way that is both precise and marks out only dogs (and not, say, wolves, coyotes or foxes). Children have concepts of dogs, but they do not have definitions of dogs. A species concept, as opposed to a definition of what the species concept refers to, is something we can have naively, so to speak. By extension we could say that scientists can use the term without being able to define it, even by their practice, and indeed they often do. But we need basic terms in science to be consistent, or else what we say about one species might be wrongly extended to species of a different group. Such confusions in science delay understanding. Hence, scientists, and those who rely upon their work such as legislators and conservationists, try to define their terms as precisely as they can.

Second, a definition is not a concept separately from the term or name of the concept. If I can define 'human' in six different ways, for example, that doesn't mean there are six concepts of *human*. There is the one concept, defined in six different ways. (This is an oversimplification. There might be two or more concepts – say, of 'human person' and 'human organism', and so forth. But that raises questions in the philosophy of language that are not relevant to us now.) The different definitions may be in competition with each other, or they may be consistent with each other. So, the fact that there are, say several hundred 'definitions' in the scientific literature (at least!) of the term and idea of *species* doesn't mean there are several hundred concepts of *species* in the literature. There are a smaller number of *conceptions* (definite ideas) of the one concept (*species*), and each conception can itself have numerous definitions and formulations. This is not generally the case with biology – for instance, *gene* has numerous distinct meanings depending on whether you are approaching it as a molecular biologist, a Mendelian geneticist, or a journalist with the latest breathless 'there is a gene for...' story. But *species* is something biologists seem to agree is a unitary concept, even though they dispute what it is. This is the reason that there is a 'species problem'.

Third, a specification of how a concept of species applies in a particular and restricted range of cases (species concepts of orangutans, for instance) is not a reason to think this multiplies the number of species concepts. A description of many dog breeds doesn't mean there are many dog concepts. There is a singular concept *dog* which is applied to numerous varieties or 'breeds'.

Fourth and finally, even if a species conception does apply neatly to a certain group, such as crows or cats or corals, it does not follow that this definition is valid for all groups of species. Sometimes adherence to a particular definition leads scientists and philosophers to deny that anything that fails to meet the criteria is a species. I'll deal with these questions in Chapter 5.

On another tack: technical terms in science often mark out competing schools of thought, and function as *boundary markers*. If scientist *A* uses *X* in one way, and scientist *B* uses *X* in another way, and they are unable to reconcile these two senses, then very often, in the words of Wittgenstein, 'each [person] ... calls the other a fool, and a heretic'. So, scientists will compete to have their preferred definition accepted or even coin new technical terms in order to demarcate the 'good' guys (= 'those I agree with') from the 'bad' guys (= 'those who oppose my view'). It's unsurprising, since scientists are human beings and do what human beings do, including play political games in their disciplines. What I do find agreeably surprising, though, is the tendency of scientists to honestly review these terms from time to time to bring them in line with the evidence. Not many human traditions or institutions do that, and this is what marks out biology, and science in general, from armchair intuitions or 'common knowledge'.

Who Uses Species, Anyway?

The most obvious answer to the question of who uses species names, classifications and even members of species is: biologists. Well, to be fair, humans in general first, but as a term of science, *species* is made for biologists. But biology is not a single profession, and it can even be argued it is not a single topic of science. Each discipline, subdiscipline, and application of biology has a different way of treating *species*.

A primary use is to organise shelves and exhibits in museums. Like a library with a classification scheme, a natural history museum needs to arrange things so they can be found, connections made with other specimens, and new trainees (also known as graduate students) taught from them.

But classification is a means to a lot of ends. Ecologists and conservation biologists need to know what species are in a region being protected. Also, they need to be able to communicate to policy makers – politicians, bureaucrats, law enforcement – what species need protection in an area and when a member of a protected species has been illicitly poisoned or shot, for example.

Biomedical researchers use species to identify model organisms used for studying how organs and processes that we humans share work. For this reason, if a species is not closely related to us (say, an octopus), we cannot draw straightforward conclusions from studying its immune system, for instance. Mice, being much more closely related, are more useful, primates more useful still.

Botanists often need to be able to identify a difference between a species and a local variety in order to study how plants behave in the wild, and so too do mycologists with fungi, microbiologists with bacteria and algae, and so on. Diatoms play a particularly important role in soils, sea and lake floors, and the ecologies that rely upon them, and diatom specialists need to be able to determine which diatoms contribute to, among other things, ocean carbon capture or food for other ocean animals.

Molecular biologists know that the biochemistry of a cell differs between species. While some processes and structures – cell membranes, Krebs cycle, DNA transcription – varies little, other aspects of cell biology are highly species-particular.

Another use is economic. Businesses depend upon proper identification of species in agriculture, horticulture, food production and so on. Fisheries, for example, need to know what species they are catching, especially when if they overfish juveniles of a species they may deplete a species their livelihood will depend on in future seasons. If you do not know that a certain species is a juvenile (too young to breed) when it is a certain size, you will not release them back into the sea correctly. Recreational fishers also need a relatively good knowledge of their prey.

Agricultural uses of species are critical, both for producing the right (and marketable) products, and for identifying pests that interfere with them. In forestry, knowledge of companion species encourages growth of trees. And of course, locals need to know which species are poisonous (or venomous) and which are safe or good to eat. Woe betide the mushroom collector who does not know the local species!

We all use species. Or do we? I mentioned that museums need to classify to store specimens. What they actually use is a name for a *specimen*. Ecologists use species names too: as a way to either identify the role played in the ecology of an area by an identified organism, or to identify a type of organism needing to be researched to find out what its role is. And so on through all of biology and biological economics. It is, I think, the *name* that is most used. The actual species itself, if the name does name an actual species, is simply the thing that anchors whatever the researchers are interested in.

So, this raises the issue why biologists are so protective of species names and descriptions. If the name is all they need to focus on ecological roles, biochemical pathways and such, why not abandon *species*? That is what we shall look at.

A Fake Story Is Essential

One of the ways in which political science-games are played is by re-engineering history. There is a story told for half a century about the notion of species and the arrival of evolutionary theory. This was created, either deliberately or not, to give status to a particular view of species in the mid-twentieth century, particularly by two architects of the 'modern synthesis', George Gaylord Simpson, an American palaeontologist, and Ernst Walter Mayr, a German-American ornithologist. It was expanded upon in 1954 by Arthur J. Cain at Oxford, in a very influential book, *Animal Species and Their Evolution*, which was reissued several times as a teaching text. The story was a way of raising the importance of Darwin against a prior group of biologists known collectively as 'neo-Lamarckians', who had adopted non-Darwinian mechanisms of evolution. This story, the 'essentialism story', has become the staple of textbook potted histories for over 60 years, and it is false.

In philosophy, an *essence* is what makes things a kind of thing; or, as we say to undergraduates, what makes a thing a member of, part of, or instance of, a *natural kind*. According to the story, only those things which have necessary and sufficient features are members of the natural kind that is defined by those features, the 'essence' of the kind. And since philosophers have been using terms for kinds in the natural, biological, world as examples of natural kinds since, well, forever, the story went that until Darwin introduced evolutionary theory, species were thought to be natural kinds with essences, which meant that they could not evolve gradually because once a single essential character changed in an organism of one species, it was immediately to be seen as a member of a new species (or maybe one that already existed, if you adopted the Great Chain of Being).

Historian Polly (Mary P.) Winsor, now emeritus at Toronto University, is the scholar who called it the 'essentialism story'. Around the time I completed my PhD attacking that story, I got in touch with Polly and found that (of course) my conclusion was not original to me. Polly was gracious enough to cite my thesis anyway in her published attack on the same story. Here's briefly how the essentialism story goes:

> When Darwin introduced the idea of species originating from previous species by a gradual process, he undercut the older creationist consensus that species had permanent essences, an idea that began with either Plato (according to Simpson) or Aristotle (according to Mayr). Since species were no longer thought to have essential properties (like body shape, inherited constitution, and so on), Darwin caused a revolution in biological systematics and in genetics.

The notion that species were thought to have essential characters before Darwin became the standard view in biology and philosophy for over 60 years, and even persists today. I'll get into the confusions and errors later, but for now I want only to ask why this story even exists in science.

Both Simpson and Mayr had their own definitions of *species*, which they wanted to be adopted by biologists generally. In science, credit for empirical and conceptual work is the coin of the realm. The more credit you are given, the more chance you have of getting students, grants, positions, assistants and the other human resources of science. This is not meant to be a cynical comment. Individual scientists may be motivated by anything from a personal desire to show some foe they are wrong, to a desire to reach truth and serve humanity. In the end, though, the proof of the pudding is in the eating. Without credit, you become a footnote in the history of your discipline, if you are remembered at all (apart from by specialist historians of biology).

Mayr, in particular, used the essentialist story as a rhetorical and partisan weapon to attack those who disputed the Modern Synthesis that he and Simpson promoted. If being non-Darwinian was equivalent to being sympathetic to pre-Darwinian ideas, then you must be an essentialist if you disputed Mayr's own species concept, which he claimed to be *the* Darwinian position (spoiler: it isn't). And being sympathetic to pre-Darwinian ideas meant you were in effect a creationist. This was the kiss of professional death in evolutionary circles, and the charge was used against many who today would be seen as clearly in line with the consensus in biology. Anyone with the least training in logic or reasoning knows this fallacy as *affirming the consequent*.

In general, species concepts in the twentieth century and beyond have been used as 'tribal flags', as banners to rally the troops during battle. It's not unusual in science, but for us not-scientists it is important to remember that just because a definition of species has been adopted widely, it isn't necessarily the most universally applicable or even conceptually the most elaborated. It may merely have had better marketing.

The Philosophy of Species

Species tend to be identified because there is breeding, and the progeny look similar to their parents. The thing that puts two organisms into members of the same species is, in keeping with the etymology of the word *species*, their appearance. But this is not the cause of being a species. This is the cause of them being put into the same species. In Chapter 3, we will consider the proposed causes of species. But we do need to make some distinctions.

There are three major tasks in philosophy, and they apply to *species* as they do to everything else. As we look at species taxa and the species concept, we shall consider all three of them. So, we distinguish between the following philosophical questions:

- What is it that constitutes something being a species (their causes, or *ontology*) and what does this say of their natures?
- How is it we know, identify, discover and refer to species (the ways we know or mark out species, or *epistemology*)?
- What value do species have for us in moral and social contexts (the ethical considerations in their use or protection, and our obligations towards them, or the *axiology*)?

I hope this book will help the reader to unravel and resolve these great questions for themselves. I shall not be presenting a 'solution' to the 'species problem' since the beginnings of the twentieth century. But all philosophers of biology, let alone all biologists, ecologists and conservationists, are required by their respective guilds to either sign on to a solution or definition of species, or try to present a new one. As my friend Matt Barker has said to me, some are allowed or even encouraged to sign on to dissolution or deflation! Think of this book as a guide to that end for you to attempt a solution for yourself.

2 Classifying *Species*

Putting Species Together

It's not enough to just list the clusters in the living world. One also needs to group clusters together within larger clusters. This process is sometimes referred to as 'ordering the world', and is called *taxonomy*, from the Greek word for 'order', *taxis*. In traditional taxonomy, begun in the sixteenth and seventeenth centuries, and formalised in the eighteenth century by Carl Linnaeus, this meant that species were grouped together in groups called in Latin *genera* (that's the plural; the singular is *genus*). As a result, Linnaeus gave each species a two-part name (a *binomial*): its genus name (which always has a capital initial) and its species 'epithet' (which is always in lowercase). So, our species binomial is *Homo sapiens*; we are the species *sapiens* in the genus *Homo*. It's kind of like a street address – you have the 'general' name (the 'street') and the 'specific' name (the 'house number') (see Box 2.1)

Species are *specific*, and genera are *generic*. The terms come from the Latin for 'source, or race' (*genus*, plural *genera*) and 'appearance, form' (*species*, both singular and plural), but, as we will see later, they have a particular set of meanings in biology. The word *species* is derived from the word *specio* ('to look'), from which we derive our words *inspect*, *spectacle*, *speculate*, *aspect* and so on. It is pronounced *spek-ee-ace* in classical Latin, but *spee-sees* in biological Latin. The classical Greek equivalent is *eidos*, or 'form', from the root word *eidon*, 'to see'. In both cases the later implication is one of appearance or outward aspect. *Genus* (Greek *genos*), however, is connected to family groups (Latin *gens*) and origins (Latin *genesis*). Oh, and just by the way, '*specie*' is a form of coinage, and not the singular of species.

Box 2.1 Groups within groups

Think of chapter one of James Joyce's *A Portrait of the Artist as a Young Man*.
A school student, Stephen Dedalus, the protagonist, gives his location and
situation in the flyleaf of his geography book:

Stephen Dedalus	(personal or first name, and family name)
Class of Elements	(subject)
Clongowes Wood College	(school)
Sallins	(locale)
County Kildare	(county)
Ireland	(country)
Europe	(region)
The World	(planet)
The Universe	(Everything)

I typically added the year level of the class when I did this as a boy. Dedalus
clearly wasn't bored enough.

Likewise, in the system that was adopted in biology early on, each species has
a similar location in the world:

sapiens	Species epithet, or name
Homo	Genus
Hominini	Tribe
Homininae	Subfamily
Hominidae	Family
Similiformes	Infraorder
Haplorhini	Suborder
Primates	Order
Mammalia	Class
Chordata	Phylum
Animalia	Kingdom
Eukaryota	Domain
Life	(Tree of life; Empire)
Nature	Everything

Genera are as multiply defined as species are in biological practice.
Taxonomists often talk about two distinct tendencies in practical biological
classification: 'lumpers' and 'splitters'. Lumpers will group many things

together based on a few discriminating criteria. Splitters will split them into many groups based on many criteria. So, a genus could include many species or few, depending on how fine-grained your discriminating criteria are. In fact, a genus can even include a single species (be *monotypic*) if the taxonomist thinks it is different enough from its nearest relatives. The trouble is that without publicly shared and preferably objective criteria for both gathering organisms into species and gathering species into genera, this is all very much a matter of personal preferences and social conventions. In short, what the binomial tells you is as much about the scientists as it is the things they study. But, since biologists actually know things about the organisms they study, this is not the same thing as saying that species are unreal or 'mere' conventions.

In recent decades (since the 1950s or so; I was born in the 1950s, so it counts as 'recent', okay?) a new approach to taxonomy has become widely adopted, although not universally or without controversy. Species are still usually described, but there are no genera (or families or even kingdoms) in the system. Instead, taxonomy, or *systematics* as it is also called, looks to visualise as much of the evolutionary relationships as it can, and the only 'real' or 'objective' *taxa* (singular, *taxon*; the terms are later back-formations from 'taxonomy' and represent the things taxonomy taxonomises) are those in which the evolutionary 'tree' has been cut just once. This is called *phylogenetic systematics* by its founder, an East Berlin Cold War era entomologist named Willi Hennig, and is often referred to as 'cladism', from the Greek word for a branch, *klados*.

Clades, by definition, include all descendants of a common ancestor, and a group is defined as being all those species that share the same last common ancestor of at least two members of the clade. Think of a family that is defined as having a single particular great grandmother – if you can define the family as 'all the people descended from the maternal great grandmother of Jenny and Jerry', then even if you don't know all the descendants of that person, and even if Jenny and Jerry aren't closely related, you have specified how to tell those people are 'part of' the family as the genealogy reports come in. In brief, a real group or taxon is held by cladists to be an entire evolutionary branch. These are the 'clades', formed from a single cut of the evolutionary tree (for which read: single ancestral taxon), and the adjective for clades is that they are *monophyletic* (a single tribe or race).

This has an interesting implication. If a primate is some organism that is in the clade specified by the *last common ancestor* (LCA, sometimes called the *cenancestor*) of, say, a howler monkey (in South America) and a macaque monkey (in Eurasia), then the group 'primates' includes humans, as we also share that ancestor. Moreover, if we specify a smaller subclade, African Great Apes, as being the clade that includes all the descendants of the LCA of, say, gorillas and gibbons, then humans are also African Great Apes. Not 'are included with them', we *are* a species of them. Arguments with traditionalists who prefer Linnaeus or his successors proceed apace. Likewise, birds must be included with dinosaurs and reptiles, and so they *are* dinosaurs and reptiles (although if they are not, then 'reptile' has no real meaning, according to cladism). Some people are very unhappy with this. But then, people were also unhappy when Linnaeus classified humans in the same genus with apes, and together with monkeys, to define the order 'primates'. Some people just don't like change, especially when they have invested so much of their lives learning the old ways.

How Many Species?

It's a simple enough question: how many species are there in the world? Various figures abound, ranging from a few tens of thousands to around 11 million (or even, if you include all single-celled organisms, around one trillion). But even though these are only estimates of the total number of known and unknown species, why is there such variation in the estimates? Could there be even more? It all depends on ... on what? Well, it depends on what you mean by 'species', and 'what-you-mean-by' is meat and drink to a philosopher. Do we include bacteria, amoebas and other single-celled organisms? Fungi? How about lichens, which are a mutual symbiosis of fungi, and one or more algae or bacteria and yeast? Viruses? And so on.

But why do biologists even use the idea of species? Especially since, as science author Richard Conniff documented so clearly in his 2011 book *The Species Seekers*, it is a dangerous business to find new species. Conniff maintains a 'wall of the dead' on his blog *strange behaviors* commemorating those naturalists who have died in pursuit of species. A friend of mine, a coleopterist (beetle specialist), was once washed away in a flooded river in Puerto Rico, losing part of his leg muscle to infection. One reason I am a philosopher is because of the comparative safety of my sofa. Few philosophers die discovering new ideas, although it has happened; the founder of logical positivism Moritz Schlick was shot in Vienna in 1936 by a disgruntled student (who then became a cause célèbre for the Nazi movement but who was probably suffering schizophrenia), and some philosophers may possibly have died through boredom. Species, though, not so safe.

The difficulty in finding, describing, naming and then studying living things (especially beetles, of which J. B. S. Haldane said repeatedly that God was inordinately fond, since there are at least 330,000 known beetle species) suggests that *species* is not a concept that biologists treat lightly. But what is it to be a species? There are multiple accounts of how it is that species come to be, and equally many on how they stay species. We will look at the coming-to-be question in the next chapter. For now, let's focus on what it means to be a species; that is, why they are stable. This is sometimes called the functionalist approach to species: what are the causes that function to maintain species?

Reproduction

Each of us has many ideas given to us in our education, our social experience and our reading. If you polled school students in years 11 and 12, my bet is that if asked to say what a species is, they would say a species is a population of living things that can breed among themselves and cannot interbreed with other populations, or words to that effect, since this is what is in the textbooks.

The very first conceptions of living species, even before the term had a technical meaning in natural history (early biology and geology), were based on several observations: that progeny resemble their parents; that progeny are caused by sexual reproduction (in animals, anyway); and that mating doesn't (often) occur between different 'forms'. There was no set term for them – genus, species, family, same-generation, same-form, and so on were all used (in Greek or Latin, mostly, but of course all cultures have notions of 'kinds' of living things). We'll get back to that, but for now, let's look at the most commonly taught definition of what a species is, the so-called *biological species concept* (BSC).

The BSC is based upon the ability of organisms to reproduce via sex. Two organisms are in the same species (that is, they are conspecific) if they are able to have fertile progeny with each other (or, if they are both of the same sex, with the other's parent of the relevant sex). If there is a definition that really resonates with biologists of big things (plants and animals) it is this one. Although the most famous definition is from Mayr, in an influential book he published in 1942, *Systematics and the Origin of Species from the Viewpoint of a Zoologist*, he was by no means the first to propose it. However, his advocacy from then on ensured that it would be the version that appeared first in every evolutionary textbook, then in genetics, and eventually in the textbooks used to teach school students. Mayr died in 2005 at 100, and he was publishing until his death, which meant that he outlived nearly all his critics. In effect, he quite literally had the last word.

Mayr defined the BSC as follows:

> A species consists of a group of populations which replace each other geographically or ecologically and of which the neighboring ones inter-grade or interbreed wherever they are in contact or which are potentially capable of doing so (with one or more of the populations) in those cases where contact is prevented by geographical or ecological barriers. Or shorter: Species are groups of actually or potentially interbreeding natural populations, which are reproductively isolated from other such groups. (*Systematics*, p. 120)

By 'interbreeding', Mayr means not only the ability to have children (progeny, as we say in biology) but to have *fertile* progeny, and not, say, mules. Species are populations that can give rise to new generations indefinitely.

However, science is not based upon a single person's views. The Royal Society, the first founded science academy that still exists (the Academy of Lynxes, which could have been a competitor, was closed for three centuries until it was re-established in 1847 by the Vatican), has as its motto *Nullius in Verba*, which means 'take nobody's word'. Critics of the BSC started early (ecologist Paul Ehrlich was one of the first) and persist, as do its defenders. One criticism is that it is not operational. In science, an idea is made 'operational' by having measurement techniques that apply either to data gathering or experimental testing. So, unless one can observe many reproductive events in a possible species, or statistically measure their ability to breed successfully, you cannot be sure that species is, in fact, a species, according to the BSC. In fact, you cannot even know if isolated populations are the same biological species. To make matters even worse, there is a non-zero chance that populations isolated for millions of years can successfully breed together. An example of this is two species of sea urchins (genus *Echinometra*) which have been separated for at least three million years by the rise of the Panama Isthmus, but which are almost as fertile with each other as they are among their own species.

A second issue with the BSC is that it is arbitrary. Arbitrary concepts and practices in science are often regarded as misleading at best and downright unscientific at worst. In the case of species, rates of gene flow between different animal and plant species vary from none to a middling amount. Under the BSC, there has to be some threshold below which genetic exchange is not enough to make two populations a single species.

There is always a possibility that if two species are close enough evolutionarily, they will share enough developmental machinery to successfully hybridise, but even then, only a small fraction of DNA may find its way back into one or both of the parental populations. What threshold should count as being within-species and what between-species? Any choice of gene-flow rate is either arbitrary, or if it can be measured experimentally or in the field may apply only to that species-pair or an individual species. Methods of comparing genes are discussed below under 'Magical Molecules', but basically, they rely upon the similarities of genes. Note that these methods do not rely upon the existence of sex, but upon the accrual of differences in gene sequences. Hence, this raises a third issue.

If the BSC is a universal definition of what it is to be a species, then any kind of organism that does not mate when reproducing, an *asexual species* (or unisexual species), is not really a species, but something else. They are called 'parthenogens' for metazoans (animals), 'apomicts' for metaphytes (fungi and plants), 'agamospecies' (usually for single-celled or very small organisms), or 'quasispecies' for viruses (a cluster of genome sequences in some measurement space). Since large parts, if not the majority, of the evolutionary tree (more on this later) don't reproduce sexually, that would mean most life doesn't come in species, which is odd, because we still name them as if they were. The realisation that most of life is asexual came late; as recently as the 1950s, proponents of the BSC could say with Mayr that asexual species are rare and aberrant. But we can't say this now. We are forced to realise that life comes in clusters, and that if we use 'species' to denote these clusters, then the BSC is not a universal species definition.

The BSC has several variations based on mate recognition systems (that is, members of the same species recognise each other as mates), genetic compatibility (genetic material is able to recombine with the DNA of another individual), ecological niche fitness (where hybrids between species have lowered fitness in an environment; this means that, over several generations, variant forms have more or fewer total descendants. More is higher, and fewer is lower), and so on, but one thing that they all have in common is that they are only identifiable at a given time period (how thick is the slice though? 100 years? 1,000? A million?), such as recent history, and not across longer periods. Over longer periods, though, a [sexual] species is less stable – it may split, or merge, or diversify into what biologists call a 'species complex', where populations vary markedly but there is continuous introgression.

Evolving Species

It has become a bit of a cliché: Darwin changed everything in biology and our self-image. And he did change many things, but the idea of evolution had only a slight effect on how biologists understood species, at least for the first half century after the *Origin* was published in 1859. Some declared that evolution meant there *were* no species, a view that was around for at least 50 years before Darwin published the *Origin*. Others declared that species were both the units and the outcomes of evolution. After the development of Mendelian genetics around 1900, though, it was the geneticists who revitalised the idea of defining species (see Chapter 4), and as genetics and evolutionary biology were slowly stitched together to form a single garment, the issue became somewhat urgent. Theodosius Dobzhansky, a Ukrainian geneticist who emigrated to America in

1927, devised his own notion of species just before Mayr, in 1935. It is now known as the *evolutionary species concept*, which he defined as follows:

> Considered dynamically, the species represents that stage of evolutionary divergence, at which the once actually or potentially interbreeding array of forms becomes segregated into two or more separate arrays which are physiologically incapable of interbreeding. ('Critique', p. 357)

Mayr thought Dobzhansky's definition was insufficient, since it was about *stages in species evolution* and not about *what was a species*. Nevertheless, evolutionary concepts, which focused on the change over time of lineages of organisms, became and remain a definition with appeal for many biologists. However, this version, abbreviated as the ESC, still requires an understanding of when one lineage is isolated from others. Herpetologist Kevin de Queiroz at the Smithsonian has a diagram (Figure 2.1) that sets up the key points.

Species criteria (labelled SC) are an arbitrary list of the things that cause a species to split that may occur in one case but need not in every case, and they may be quite diverse. I say 'arbitrary' because biologists can rely upon infertility, or mate recognition criteria, or the accrual of characters like colour, form, geography, ecological role or behaviour, or just about anything else that they deem worthwhile – the list is not forced on them by theory. The idea at the core of the ESC is that species are largely homogeneous until they begin to split and after they have split, but that during the splitting process, the lines leading to the two new split lineages can look like subspecies, varieties or local variations, largely because they will routinely share DNA through mating at that stage when in contact. The choice for the biologist as to when to say 'enough' has occurred to make two new species itself varies case by case.

Part of the difficulty with evolutionary views of species is that at best a species ascription is only temporary and provisional. Maybe the two lines of populations will rejoin in the future. Some of these conceptions rely on such vague terms as 'evolutionary fate' or 'evolved differences'. However, lacking a forward-looking chronoscope (a 'predictascope™', coming to a timeline near you), we cannot say with any degree of certainty that two species might be permanently distinct. Since closely related species share their cenancestor's fertilisation and developmental systems, there is a decreasing but non-zero chance they will hybridise when in contact: lions and tigers are separated

Species criteria

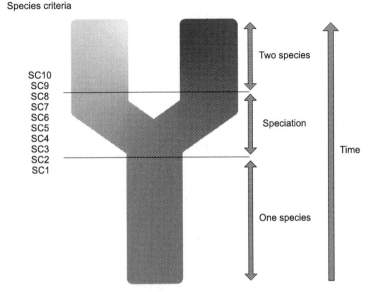

Figure 2.1   Evolutionary species.
Evolutionary species conceptions can give different species counts as they reflect changes in the lineages and their populations' makeup. Each set of species criteria (numbered 1 to 10) applied may give one, or two or an indeterminate number of species. During speciation, some variations will be considered as subspecies (varieties, geographic populations, and so on), until they meet the criteria.

by around seven million years of independent evolution, and yet they can interbreed to form tigons, ligers and a whole host of secondary crossbreeds. The longer the separation, in animals at least, the less likely that the progeny will be fertile or healthy. Plants, on the other hand, hybridise much more frequently. This raises the question of what counts as distinct evolutionary paths. So, the evolutionary species concept is not so operational either.

One approach to systematics (the study of the diversity of organisms) is called *evolutionary systematics*, and it focuses on the diversity of traits rather than just species. Here it is the evolutionary differences, and not the similarities, that are the focus of classification. However, while this approach focuses on changes between species, and higher groups, it also focuses on within-species variation as well. Population genetics, as noted below, involves a fusion of what Mayr called 'population thinking', which is another term for statistical thinking about units of inheritance, and Mendelian genetics. Early evolutionary geneticists saw species as variable genetic lineages, represented by population curves for variation.

This approach often also makes arbitrary choices about what similarities are to be used by the taxonomist. Similarity is based on the differences that strike the taxonomist as important (see Chapter 5). As a result, there is constant debate over whether this or that difference is truly significant. Some really spectacular disputes in science result from this and related debates (systematists can be a quarrelsome lot, quite unlike, say, philosophers). Some of the debates are about which traits are informative to the taxonomy; but some are more theoretical: should we count useful traits or not, and are there adaptive grades or states of functional adaptations? Should all traits used to identify be characters that are the same part inherited from a common ancestor or not? And so on.

The central notion of evolutionary species conceptions is of a *lineage*, most usually of populations or overall collections of populations (*metapopulations*) of organisms. Lineages are, according to Dobzhansky, parent–progeny chains. However, lineages of genes, body forms and even adaptive traits have also been used at times, and they very often do not match the lineages of a species' populations. Gene trees are not often species trees, or at least not on their own. Recently, de Queiroz proposed his General Lineage Concept, which he later developed into the Unified Species Concept. Under this notion, a species is a separate lineage of populations (which themselves are built out of genetic, familial and cell-level lineages). One issue, though, is why some lineages are thought to be species when others are not. For example, there are lineages of populations within species (think of ethnic groups in humans, or the Alaskan and Californian populations of sea otters).

Phylogenies

De Queiroz is also one of those biologists involved in formulating the PhyloCode, an attempt to redefine systematics along cladistic lines (see below), but which retains one rank only: species. This is seen as progress by many systematists, but the failure to abandon all ranks is seen by some (including me) as an inconsistency in the principles applied. De Queiroz doesn't think species *is* a rank, but that it is instead something that can be conditionally defined by secondary characteristics, none of which are universal, including reproductive compatibility.

The way in which PhyloCode, and for that matter, most of systematics nowadays, structures its classifications is by tree diagrams known as *cladograms* (which come in a number of varieties we won't bother with here). These diagrams are referred to as *phylogenies* ('phylogeny' derives from 'origin of the tribe' in Greek. Biologists sometimes criticise *philosophers* for abstruse terminologies) and are – very roughly – evolutionary trees. A phylogeny is like a genealogical family tree, only for species and larger groups. For our purposes, we can think of them as evolutionary trees.

Homologies – the characters that are 'the same' – are the sole criterion for classifying species under the method called *phylogenetic systematics* (or 'cladistics'). There are several kinds and interpretations of homology, which we won't go into here, but the dictionary definition of a homology as 'similarity due to descent from a common ancestor' (summarising Merriam-Webster) is incorrect. For a start, homologous organs or structures (*homologs*) may not resemble each other much in different species. Also, similar-looking structures may have evolved convergently, in what is sometimes called 'parallel evolution' or *homoplasy* (-plasy comes from the same root as 'plastic' and means much the same: something that can be easily modified). Moreover, the dictionary definition mistakes the *explanation* of homology (descent from ancestral structures) for the homological relationship itself. Homologies were defined well before Darwin's theory of evolution was published (very early on known as 'analogues' or 'affinities'). According to Richard Owen, who coined the term in 1843, a homolog is '[t]he same organ in different animals under every variety of form or function'. In other words, homology is the relationship between parts of different organisms no matter whether they do the same things or look the same. A rib is a rib, even if it is part of a turtle's shell or a snake's body. Homologs just have to be the same *part*, which is worked out by its arrangement with other parts like skulls or hearts (or cells or genes – these days homologs go all the way down deep into the molecular scale).

Cladistics doesn't, by itself, have a notion of species. The East German entomologist who developed the methodology during the Cold War, Willi Hennig, adopted

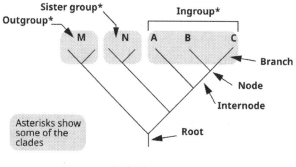

A cladogram

Figure 2.2 A cladogram (branch diagram).

Mayr's reproductive notion of a species, and treated specimens of species as the terminal nodes in the tree diagram (Figure 2.2). What was novel about his approach was the explicit reliance on homologs to identify *monophyly* (recall this is where the evolutionary tree is cut once and only once). Only monophyletic groups were considered 'real', 'natural' or 'true' classifications. Hennig also renamed homologs to *apomorphies*, from the Greek 'forms from'. He was really enthusiastic about Greek-derived novel terms. An apomorphy is just a homolog that has changed since the last common ancestor of the organisms.

Of course, since many 'good' species are not monophyletic, this led to a series of 'phylogenetic species concepts', which redefined *species* as a clade, or at least as a monophyletic group in the analysis of the data on homologous characters. Here are some examples:

> A species is the smallest diagnosable cluster of individual organisms within which there is a parental pattern of ancestry and descent (Cracraft, 'Species concepts')
>
> A geographically constrained group of individuals with some unique apomorphous characters, the unit of evolutionary significance (Rosen, 'Fishes' – apomorphous here means traits that can be identified as homologs)

... the smallest aggregation of (sexual) populations or (asexual) lineages diagnosable by a unique combination of character traits (Wheeler and Platnick, 'Phylogenetic species concept')

This is where it gets messy, as 'ordinary' (good or valid) species in the Linnaean scheme, where species is the lowest persistent unit, often have phylogenies (separate genealogies) within them (that is, there will be clades 'inside' the usual species), and so adopting a phylogenetic definition means that 'good' species must often be dissolved into many smaller 'phylospecies', causing what some call *taxonomic inflation*. This becomes a critical problem in conservation biology, and we will return to it later.

So either

- species are supposed to be monophyletic, in which case most presently defined species must be revised, causing taxonomic inflation, or
- species are not monophyletic, in which case they stand out as the one type of not-monophyletic taxon that is 'natural' when otherwise only clades are natural.

This problem remains unresolved in systematics. When the monophyletic definition is applied, the number of species can triple or more. This makes conservation very difficult, when it relies upon species as the 'unit' of conservation, for example. An example: there is supposedly one gorilla species: *Gorilla gorilla*. Recently, however, it has been split into four subspecies of two species: *G. gorilla gorilla* and *G. gorilla dielhi*, and *G. beringei graueri* and *G. beringei beringei*. Now is this two species, or four species, or one species with four subpopulations? If a choice is made, does that also determine the conservation priorities? There is only a small amount of money and time allocated to protecting endangered species, so what to do? This is more than an abstract question.

Magical Molecules

We are all broadly aware of the role molecular biology plays these days – at least the molecule called deoxyribonucleic acid, or DNA. However, the use of molecules to differentiate species is quite recent. The *genome* of a species is often thought to contain markers that are unique to all its members. These views are grouped together as *Genetic Species Concepts*. Here are some examples:

> A Mendelian population is a reproductive community of sexual and cross-fertilizing individuals which share in a common gene pool. . . . The biological species is the largest and most exclusive Mendelian population. (Dobzhansky, 'Mendelian populations')
>
> . . . a group of genetically compatible interbreeding natural populations that is genetically isolated from other such groups (Baker and Bradley, 'Speciation in mammals')

Before current techniques for rapid genomic sequencing became available, several shortcuts were used. The first technique used was DNA hybridisation, a method that involved 'cutting up' DNA molecules into fragments, heating the solution in order for the two strands of each of these DNA fragments to separate from each other and then allowing these single-stranded fragments to align with those of the compared species' DNA. The amount of matching up gave a figure of 'sequence similarity'. Today, we can identify gene sequences at the base-pair level (of adenine, thymine, guanine and cytosine, or ATGC) and be much more precise. Okay then, how much similarity applies between species? Well, it depends on the organisms. For viruses, it's around 80%. For hookworms, it's between 93% and 99.1%. For cattle lungworm, it's around 50–76%, and so on. And this all depends on already knowing the species. So, the choice of a difference threshold in one group of organisms cannot be generalised to all organisms. For example, if we used the hookworm threshold, then chimps, bonobos and humans are the 'same' species as we share some 98.7% of our DNA. The lungworm threshold means all primates could be considered the same species. Clearly this doesn't work.

Later, a version of sequence similarity called *DNA barcoding* used a shared 'location' (*locus*) on the genetic material of one of the organelles ('little organs') in cells that have them. The gene in animals is called *cytochrome c oxidase* I (a metabolic gene, *COX1* or *CO1*; gene sequence names are italicised, and their proteins are roman in print) from mitochondria; in plants, it is the *rbcL* gene from chloroplasts which is involved in taking in carbon; in fungi, the *ITSrRNA* sequence; in single-celled eukaryotes, the *18SrRNA* sequence; and in single-celled prokaryotes, *16SrRNA* is used. Again, the thresholds range from 60% to 97% for species identification.

The rationale for molecular/genetic concepts is straightforward. Genes are (supposed to be) what make organisms what they are; therefore, genes can diagnose species, a form of molecular essentialism. However, the assumptions that underlie this rationale are questionable. DNA is not a magic molecule – it varies within species as well as between them, and while there may be some species that are low in genetic diversity and have 'speciation genes' as they have been called, not all species are likely to have them. Moreover, the phenotype (body structure) of a typical member of a species is affected by non-genetic factors, such as the availability of certain foods during development, the environment in general, the conditions of development (in the womb, egg or seedling) and maturation. There is a broad consensus that genes interact with the environment, including the parental environment, to form the adult organism, expressed with the symbolism $G \times E = P$ (genes and the environment result in the phenotype). This suggests that while genes may be necessary to understanding species, they are not enough.

'Genetic' also has several meanings in biology, depending on which field you are working in. It can mean the heritable 'factors' of Gregor Mendel (which early Mendelians later called 'genes'), which are effectively just recurring traits passed on to progeny. This gave rise to a field called *population genetics* once Mendel and Darwin were being synthesised into one science, and population genetics ('popgen') was often taken to be the core of evolutionary biology and of systematics. For many it still is. As a result, genetic conceptions of species are still considered viable. Ironically, as we shall see later, Mendelian genetics gave rise to the so-called 'species problem' in the first instance.

But 'genetic' can also mean an expressed sequence of DNA molecules. Or it can mean the expressed sequence as 'edited' by RNA molecules. And on it goes. 'Gene' is as ambiguous a word in biology as species. Using genes to define species is to use the unclear to define the ambiguous.

Some genetic species conceptions involve genes that cause speciation (species differentiation; see Chapter 3). These 'speciation genes' are causes of the differentiation of populations or lineages, and so are not arbitrary. However, while some genes may be causes of some speciation events, there don't appear to be universal genes, or types of genes, that are causes of all species.

Other Sorts of Kinds of Species

There are several other sorts of species that are proposed. For now, though, I want to mention the *ecological species* (*ecospecies*) notion. It was proposed because of the discovery early in the twentieth century that some plants can present the same features over time even if the underlying genetics is diverse or variable. For example, eucalypts in Australia include around 934 species of *Angophora*, *Corymbia* and *Eucalyptus*; some are not genetically homogeneous, and yet their features seem to be maintained by the environmental conditions in which they grow, such as on a riverbank (called 'river gums'). Moreover, eucalyptus species can often hybridise even between genera, although hybrids between more closely related species do better.

A similar situation applies to American oaks of the genus *Quercus*. American palaeontologist Leigh Van Valen offered, in 1976, a definition of ecological species.

A species is a lineage (or a closely related set of lineages) which occupies an adaptive zone minimally different from that of any other lineage in its range and which evolves separately from all lineages outside its range.

The defining concept used here is 'adaptive zone', a concept introduced by another American palaeontologist, George Gaylord Simpson (who also introduced the notion of 'lineage'). Van Valen used the term to mean a resource space, along with interactions with predators and prey, and so forth, and like Simpson thought that adaptive zones are not fixed but themselves evolve. But the notion can be abused. Simpson's contemporary, Julian Huxley (another eugenicist as well as a zoologist, and Aldous Huxley's older brother) even went so far, in a fit of human hubris, to suggest there was an adaptive zone, a grade or rank of organisation, of intelligence, *Psychozoa*. Unsurprisingly, only humans were known to inhabit this grade. (This human exceptionalism seems to happen a lot in biology, and in philosophy. But all species are unique, or they wouldn't be species.)

Again, we run up against conceptual and definitional vagueness. As with terms like 'niche' in ecology, there appear to be no clear criteria for adaptive zone ascription. It isn't that the idea is purely subjective or conventional, but rather that there seems to be something recognisable about adaptive zones/grades which practitioners know but cannot express precisely. Adaptation is due to a plurality of fitness-enhancing traits, and the evolutionary jury is still out on intelligence. Time will tell.

Classification

Species play a special role in classification, but why is that important? There aren't, for instance, species of items in a library, or species of hardware in a store. Most may be books or fasteners, but there are a wide variety of other types of items in both places. One can classify these objects without a base unit, so why do biologists think they need such a rank? In part, species matter for taxonomists because there are different forms of body type, called the *phenotype* in biology. These can be seen and studied, of course, and under the Essentialist Story, this became known as the *Morphological Species Concept*. It was basically the 'definition' of museums and of bird watchers using field guides.

> Species are the smallest groups that are consistently and persistently distinct, and distinguishable by ordinary means. (Cronquist, 'Once again')

Except it wasn't really a definition, and a type of form did not imply essence, just typicality. The Morphological Species Concept was an invention of Mayr and others to denigrate the ways in which taxonomy and systematics had been done prior to their 'New Systematics' of the 1940s and afterwards. The irony was that these new 'Darwinian' notions did not affect the ways species were described all that much until the molecular revolution of the 1970s and 1980s. Every biologist, and anatomist, that ever described and named a species considered the form, or morphology, of the specimens being studied. They still do, and so did Darwin himself. The so-called Morphological Concept never existed except as a set of practices.

Morphology played a significant role in *numerical systematics*, which was also known as 'phenetics' (from the Greek word for 'appearance', *phaeneros*), which enjoyed wide popularity in the 1960s and 1970s. Here characters of all kinds were quantified, and mathematical techniques well suited to computers (then becoming widely used by biologists) were employed to group organisms into what was called *operational taxonomic units* or OTUs. In an attempt to remove all subjective bias, though, the numerical taxonomists included any and all characters, not merely homologies, and this meant that, based on the characters used, organisms could be classified in a number of OTUs, and so OTUs could dissolve or radically change with new data – in short, they were not stable. Although OTUs were not species, they were supposed to be replacements for species (see Chapter 9).

Species by Convention or Convenience

> By the way I met the other day [John] Phillips, the Palaeontologist, & he asked me 'how do you define a species?'—I answered, 'I cannot' Whereupon he said, 'at last I have found out the only true definition, —'any form which has ever had a specific name'!. (Charles Darwin to Asa Gray, 29 November 1857)

> A species is a community, or a number of communities, whose distinctive morphological characters are, in the opinion of a competent systematist, sufficiently definite to entitle it, or them, to a specific name. (Regan, 'Organic evolution')

Some people think that species are things made up for our convenience or just brought out of our practices and 'folk knowledge'. This is sometimes called the *Conventional Species Concept*, although the philosopher Phillip Kitcher once called this the Cynical Species Concept. That is, I think, a bit unfair. It is not cynical, but a matter of acquired professional training and expertise. However, while it is not cynical, neither is it a concept. A conventional species is the product of institutional behaviours and conventions (hence the name I have given it). It is just the competent use of evidence, usually morphological but not always, to describe and name kinds of observed organisms, by those who specialise in that process and are familiar with the group within which those organisms fall. Also, except in those cases where some DNA is retrievable from fairly recent specimens, morphology is the only way to describe and name fossil species, which tends to lead to debates over whether or not similar fossils are members of a distinct or just a highly diverse species.

Are Humans One or More Species?

To make these definitions clearer to those who do not think in abstractions, I'd like to discuss how a species that many people are somewhat familiar with fares under each definition. Called *Homo sapiens* (in case you missed the irony, that's us), it is a member of the African Great Apes clade, but it is now found in all continents and climates and is about to migrate to other planets, apparently.

H. sapiens originated mostly in Africa but has since interbred with other members of the *Homo* genus: Neandertals in the Middle East and Europe, an Asian species known as Denisovans, and possibly up to three or four others. This species reproduces sexually (in the absence of artificial wombs and cloning technology and a complete lack of ethics) and is the sole remaining species of the group from which it sprang.

The extinct members of the group are *Homo habilis*, *Homo neanderthalensis*, *Homo ergaster*, the as-yet unnamed Denisovan species (it'll probably be called *H. altaiensis* once they find enough of a specimen to designate), and about six or seven others. A simplified evolutionary tree for this species is shown in Figure 2.3.

The sharing of genetic material from these Neandertals and Denisovans (at least) has led to several lineages within humans that are marked by these genes (although not necessarily showing anything superficially obvious). These mixed populations appear mostly outside of Africa, especially in Eurasia (Iberia, the middle east, and west Asia), but also in the Pacific region, such as Papua New Guinea and Melanesia. Neandertal DNA is more frequent in Europe and west Asia, while Denisovan DNA is found only in east Asia and the Pacitic. African populations have far less Neandertal and almost no Denisovan DNA.

Now, on the various definitions of what a species is, we get various parts of the evolutionary tree marked out as 'human' species. All *Homo* species are human, of course, but which ones are *us*? Starting with the traditional

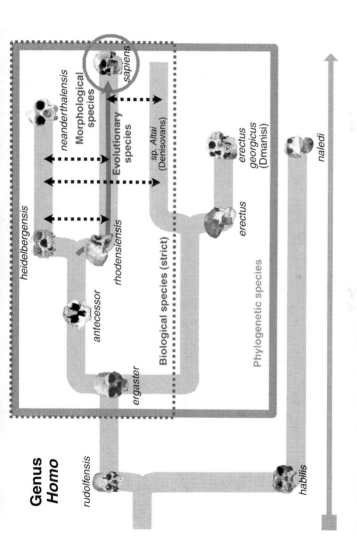

Figure 2.3 Human species under various definitions.

A tree of the genus *Homo* showing how different definitions can be applied to give wildly variable results. Dotted black arrows show genetic exchanges between species, but these are rare events. Still, some populations of non-African *H. sapiens* have a considerable genetic inheritance from both Neandertals and the Denisovan species, and probably others. This phylogeny is probably wrong and will need to be revised as more discoveries are made.

'morphological' identification, that is, the definition from Linnaeus (which was 'know thyself'; today we treat Linnaeus himself as the type specimen of humanity), just the present populations and their immediate ancestors are a species. For a time, Asians, Indigenous Americans and Africans were also thought to be separate species (the 'polygenist' claim, used by proponents of African slavery), but this is not a mistake Linnaeus made. This view of 'us' is a horizontal definition and equates the species with the terminal node of the evolutionary tree upon which we are located (the circle in Figure 2.3). Our experience is that all populations of this node can freely interbreed without problem, and so it is considered a biological species (that is, a group of reproductive communities), but this is not how we decide if it is worth a name, owing to the issues mentioned with the BSC.

If we adhere strictly to Mayr's original definition of biological species, then the fact that there is hybridisation between these various evolutionary lineages means that they must include all species with the same basic developmental system, and so a truly biological species must include all common ancestors of the groups that can interbreed (Figure 2.3, dashed line). This is wild, and it would mean that we are all in fact sub-varieties of *H. ergaster*. Or, according to the taxonomic rules, we could call all these species *Homo sapiens* (since it was the first name used by Linnaeus) with a 'subspecies' name such as *ergaster* or *antecessor* for each of the different forms.

Either way, we must, and as a matter of practical application we do, relax the original BSC so that occasional hybrid events do not destroy the integrity of the species. This is very important, because as biologists use genetic tools to identify relations between species in many organisms, they are finding increasing amount of crossbreeding. Again, though, there are variant forms of *sapiens* that are called 'archaic' or Cro-Magnon, that scientists debate whether they are species or varieties or just normal variation. Similarly with racial differences, although, despite our sensitivity to apparent differences, the amount of genetic variation in humans is well below that of most large-bodied animals. 'Race' is not a biological term of much use in biology for humans. The borders of biological species are getting harder to locate (a field referred to as species delimitation).

So perhaps we should take an evolutionary approach (Figure 2.3, solid line with arrow). Here there is a separate evolutionary fate from *H. rhodensiensis* to *H. sapiens*, but we do not know merely from cranial or skeletal features whether the *H. rhodensiensis* is directly ancestral to us. Palaeontological inferences are not that well-founded. Maybe there was a closely related species to *rhodensiensis* that was similar and from which we truly did evolve. In principle, evolutionary accounts may in fact be the reality of the underlying diversity, but the actual sequences can only be inferred from what fossil data we have, and that is generally not much. Moreover, the introgression of Denisovan and Neandertal genes suggests that to some degree, the *sapiens* lineage does not have an 'independent evolutionary fate'. Like the BSC, with the evolutionary conception we need to be slightly arbitrary in our distinguishing of the borders.

This leads rather directly to another lineage-based conception of *species*: the *phylogenetic species* or *phylospecies*. The sole criterion for phylospecies is that they are all the subsequent parts of the evolutionary tree (as reconstructed by phylogenetic systematics) from a single smallest cut of the tree. Returning to our morphological species notion, maybe that could be a phylospecies as well. But there's a problem with that. See the dotted black lines that indicate crossbreeding between [populational] lineages? This means that the morphological species *H. sapiens* is not *monophyletic*, which is to say it has multiple roots in Neandertals, Denisovans and probably several others. So, if we expand our definition of *H. sapiens* to require strict monophyly, we must include the last common ancestor of these lineages. This again takes us back to *ergaster*. But if we say something like 'Well, a small amount of gene flow between lineages is not enough to make the populations not monophyletic', then we have given away the phylospecies concept, because it holds that gene flow is what identifies something as a populational lineage, and if we can ignore some gene flow we must have a prior idea of what makes a species before we can say this is a phylospecies at all. In short, it is question-begging.

To make matters worse, as I mentioned several times, many species have *within*-species lineages, some of which are clades. These are often isolated populations (like the Alaskan and Californian sea otters), or they are *haplotypes* (as the journal *Nature* defines them, 'the genetic makeup of an individual with respect to combinations of alleles that are closely linked and tend to be inherited together'). A *haplotype group* is a group of organisms that share unique genes that have a single common ancestor. So why aren't these called species? And again, the answer is: because we already have an idea of what a species for mammals is, and the haplotype scale is too small.

So it is difficult to find any semblance of a unitary and universal definition of what makes something a species. Now that we are suitably confused and dazed about the concept of species in biology, let's get equally confused about how species (whatever they are) come to be: *speciation*.

3 Making Species

Every textbook of biology will supply a number of 'modes of speciation', the ways in which new species evolve. But the issues in dispute among the biologists themselves are rather odd. The adoption of evolutionary theory by biologists has had a great impact on how species are understood. From the idea that kinds of living beings were created and at best had devolved to localised varieties, now species were the target of a 'mechanical' or 'physiological' explanation: they came into being. And under Darwin's version of the evolutionary account (initially known as the 'development theory', since the Latin word *evolutio* means 'development'), species were made from other, allied (which means 'closely related'), species. The processes and causes of new species set up the 'species question' that Darwin and other naturalists were seeking to answer.

Darwin thought that species were formed by adaptation to novel environments, via natural selection. Others, in France and Germany, agreed, but some felt that isolation of the population from the original species was required as well. The disagreement presaged an extensive debate during the twentieth century on how species were made, which led to a subdiscipline of evolutionary biology called speciation theory. It is also known as *macroevolution*, and changes within a species are called *microevolution*, terms coined in 1927 by the Russian entomologist Iuri'i Filipchenko.

One point about these words, since many biologists and creationists alike misuse them: any evolutionary changes above the rank of species are *macro*evolutionary, not just speciation, and not just major changes in body plans, adaptive zones or post-extinction radiations. Creationists take macroevolution to mean changes that could not have taken place after the Ark; so, all cats have speciated from the single pair of cats on the Ark, but cats did not evolve from a proto-carnivore, according to them. Creationism ironically requires more and quicker evolution than Darwinism to fit into both the Ark and the time available. Many scientists tend to take macroevolution to mean something similar: changes too great to have been just simple speciation, and which require processes acting above the level of species.

Speciation was therefore hotly debated in biology from the viewpoint of the so-called modern synthesis of evolutionary theory, also sometimes called 'Darwinism', and genetics. Over time, the field settled into a mostly *geographical* classification of speciation (see Table 3.1): that is to say, whether species are formed when isolated from each other or when populations diverge in the same area to form new species.

Later in the twentieth century, as molecular biology and microbiology advanced, problems arose once the synthesis was expanded to include asexual organisms and hybridisation, as well as chromosomal (i.e., genetic structural) changes. This all led to a renewal of the species question.

The Temporal Process

Since the beginning of western thought, the view that everything changes has been a worry. Heraclitus had problems with rivers, and Aristotle spent a lot of time on essences and generation. Modern evolutionary thought holds that change is a brute fact of the living world which explains why things are as they are. However, for a very long time, the standard view was that entire species gradually changed over time to become new species. You can see examples of this idea illustrated in early twentieth-century texts. They basically look like Figure 3.1.

Chronospecies are species identified by sampling fossil variation in form, usually skeletal or shell-based, across time, divided by taxonomists into species either at an arbitrary time point, or more likely because of sampling gaps. Arguably, several of the *Homo* species are chronospecies – maybe we are just a modified *Homo ergaster* subspecies with pretensions, as I suggested earlier.

Type	Name	Definition
Time	Chronospeciation	Gradual accrual of changes leading to something different enough to be called a new species
Migration (geography)	Allopatric (dichopatric)	Barriers to mating accrue during geographic isolation from the parental population
	Parapatric	Allopatric, but bordering with and occasionally hybridising with the parental populations
	Sympatric	Distribution of fitness levels with intervening low fitness phenotypes leading to reproductive isolation
	Stasipatric	Speciation of semi-isolated populations (e.g., on mountain peaks or islands, or in lakes) within the range of a species (often by polyploidy)
Ploidy (number of chromosomes)	Polyploidy	Increases in the number of chromosomes
	Autopolyploidy	Multiplication of the normal number of chromosomes by an integer (e.g., $2n$ or $3n$)
	Allopolyploidy	Multiplication of chromosomes including some from another species
Other	Non-genetic factors	Phage or parasite transmission (e.g., *Wolbachia*; see text)
		Behavioural barriers to interbreeding
		Ecological barriers to interbreeding
	Asexual/unisexual	Conjugation in protists (see text)
		Quasispecies (see text)
		Apomixis/parthenogenesis (see text)

Table 3.1   Types of speciation processes

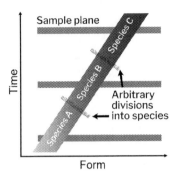

Figure 3.1 The gradualist view of speciation.
Chronospecies diverging by gradual change in a single species, or *anagenesis*. The divisions between species are arbitrary.

Obviously, this conception of species depends largely on the available data and is thus a matter of convention: where (or if a gradual change, when) is it convenient to divide the spectrum? The 'reality' of these species is that sampling the morphology (the form) of specimens from each geological layer shows measurable differences that, if these specimens had lived at the same time and place, mean they would have been called distinct species.

More recently, discussions about the rates of evolution of species and phylogenies (a phylogeny is a reconstructed relationship expressed as a tree diagram) have argued that the palaeontological approach of the earlier systematists is misleading, and that it is much better to see such 'transitions' as species at different times related in a tree format (Figure 3.2).

Now it might be that species A through C in these diagrams actually did evolve by a process of gradual change, in which case this would be speciation known as *anagenesis* or the evolution of new species without splitting (from 'origin up' in Greek). But it might also, given the same data, be a branching tree like Figure 3.2 (*cladogenesis*, speciation by splitting, or 'branching origin' in Greek). Or it might be some combination of these (a recent review of human evolution adds *reticulation* or joining networks). To argue that it was any one of these, the researcher needs to know what processes were happening at the

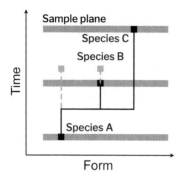

Figure 3.2   The punctuated or episodic view of speciation.
Chronospecies as nodes in lineages that diverge through abrupt evolution, or
cladogenesis. The divisions are less arbitrary owing to speciation events.

time. For example, it is very unlikely that an entire species of any extensive
range would be subjected to either the same natural selection pressures, or the
same genetic mixtures, although a species that consists of one population in
a confined area might. This means we have to be able to rule out, with some
confidence, the other more likely processes (so far as we know) in order to be
comfortable with anagenetic accounts. Since we don't have a lot of information
about the past situation for any specimen, instead researchers will use an evolutionary model with what look like reasonable assumptions. This runs into trouble
when the models – which are hypothetical – are used as evidence, since the end-
result is circular ('the model says this happened, proving that the model is right').
This is a widespread difficulty in thinking about the past. Sometimes we make
assumptions that are good, such as the laws of physics not changing rapidly and
arbitrarily over time. But if we assume gradual speciation in order to prove
gradual speciation occurred, we are just putting our thumbs on the scales.

An alternative approach is to take seriously the fragmentary data that we do
have and suppose that the evolution that occurs is not anagenetic but episodic,
that is, species form 'suddenly' (in geological terms), which can mean many or
even thousands of years. Stephen Jay Gould and Niles Eldredge called this
'punctuated equilibrium' (PE) – that species form quickly and then remain

pretty well constant thereafter. The idea is that species change rapidly in response to several possible processes, including those involving climate, geology, genetic structure, population size and, just generally, chance. But this, too, can be circular. Absence of evidence (of gradual evolution) is not evidence of absence (of gradual evolution) either.

The jury is still, after some 40 years, out on this debate, but when PE was proposed it caused a furore among evolutionary biologists that has not entirely settled even now. It seems to me that the likelihood is that some do evolve rapidly, and others do evolve gradually, and these events can happen in the same lineage. There may be no general rule about how species form, and this makes sense to me, at any rate: biology is mostly contingent and local, and each case will have its own history.

Consider the extinction of all large-bodied animals when the bolide hit 66.5 million years ago. A single cause explanation makes sense here. But the reasons why each individual small-bodied species survived will not be the same for all species. So the same is true of the evolution of new species. Some may follow one pattern or the other, or a mixture, or even something we haven't thought of yet. But assuming the explanation in interpreting the information given to us to support the explanation should never be done.

Species in Space

Geography is a strange way to think of how species form, at first. But it is really based upon population genetics, and the fact that populations occupy a geographical range of some sort. A key issue in speciation is migration, which is to say, transfer of genetic material from one population to another as individual organisms migrate between them and mate. If genetic material spreads into another population, then the receiving population may have more genetic resources to deal with environmental challenges than it did before. What happens next is due at least in part to the geography of the species itself – do they merge and share all genetic differences, or do they remain distinct even when they share a territory, and possibly food sources? There are overlapping populations of crows in Europe that have maintained their 'hybrid zone' for at least centuries without changing either of the contributing species. One caveat though: this is about sexually reproducing species, or at least populations where two individuals contribute genetic material to the offspring. We will get to asexual organisms later.

If a subpopulation becomes isolated from the main population for long enough, and if it has a somewhat different sampling of the gene pool from that original population, then random events (an organism with type *y* genes might mate with an organism with type *x* genes which happen to work together to create a new phenotype) will result in a different overall population than the two parent species. Given enough time, evolutionary changes due to mutation and random shuffling of genetic variants in reproduction will lead to an increase in the difficulty of interbreeding if the two come back into contact.

In short, these 'accidental' changes cause the reduced fertility, rather than reduced fertility causing new species due to hybrid incompatibilities. The more infertile the hybrids, the more likely the isolated population will become a species. However, infertility is not absolute: it can mean that the hybrids are fully infertile, or it can mean that they have lowered fitness due to developmental problems (as in ligers and tigons), or that they are not well adapted to the environmental conditions of either population. For example, even fully fertile hybrids may fail, because they do not match the conditions of either parent species. There are hybrid lizards between dark forms and light forms of sister species that live, respectively, on dark rocks or light sand. Hybrids are neither dark enough nor light enough to camouflage them against predators, and so they get eaten. Similar things have been seen with birds, fishes, moths and butterflies.

In the end, for two populations to become a species, what counts is not the fertility of hybrids, but the long-term fitness of hybrid progeny. Since fitness is a measure of long-term success, natural selection doesn't make species so much as keep them distinct once they arise.

Populations, Ranges and Migration

It is, wrote the Roman poet Horace, fit and proper to die for one's homeland. The Latin word he used for homeland was 'patria' (*dulce et decorum est pro patria mori*), and the word has entered into biology as the root for exactly that (compare 'patriot'; *patria* is a loan word from classical Greek). Unlike Horace's slogan, though, it applies more to living than dying. It would be nice if we humans could attempt to live for our homelands rather than die for them, but that's for another discourse.

There is a cluster of terms used by biologists to describe where organisms live or grow (Figure 3.3), and they are based on the Greek term *patris*: *sympatric, allopatric, parapatric, peripatric, stasipatric* and *dichopatric*. This flock of technical terms is confusing to the newcomer (and to some biologists), but there is a kind of logic – as much as in the evolution of any technical jargon – that will make it clearer, and at the same time allow us to set up the alternative views on the fundamental evolutionary process of common descent: speciation.

What is speciation? It is the formation of new species. Most of the time in most groups of organisms, species are formed when one splits into two, rarely more than two. Sometimes species form by hybridising, or crossbreeding. Two organisms, or two populations, or two varieties, species or taxa, can live in relation to each other in two ways, geographically speaking. They can live in the same region (*sym-* meaning 'together', plus *patris*, gives sympatry), or they can live in different regions (*allo-* meaning 'other', plus *patris* gives allopatry).

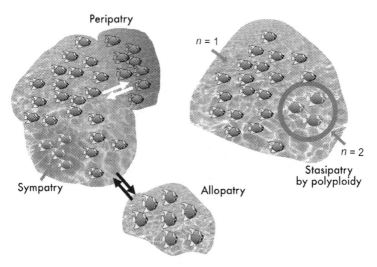

Figure 3.3   Speciation by geography.
Speciation modes categorised by geography. I have chosen fish here because it shows that such issues can arise even when there is freedom of movement by individual organisms.

So, if new species split from a common ancestor species, they might do it in these two ways. One is how Darwin thought of it, referred to today as *sympatric speciation*. He believed that varieties within a species became adapted to local conditions in different ways. A frog might adapt to eating one sort of fly, and another member of its species, called a *conspecific*, might adapt to eating a different one. If the food were found in slightly different conditions, or at different times (one might fly during the heat of the day, while another might only fly at dusk), then the varieties might adapt to those times and conditions. Eventually their adaptations could incidentally cause so much difference in behaviour or structure or climatic preference that they would no longer routinely mate.

It is important to note about sympatric speciation, both in Darwin's mind and in the minds of those who today agree with him, that natural selection causes the adaptations, but not the speciation itself. Most biological notions of species involve sexual isolation or the inability to, or reluctance to naturally, breed together. Darwin thought that the reluctance to interbreed would follow as a by-product of adaptation to these environmental conditions.

The other way, *allopatric speciation*, proposed by a German writer named Moritz Wagner in Darwin's time (although a French writer, Pierre Trémaux, got there first by a few years), and with whom Darwin corresponded, involved populations of a species becoming isolated from each other (which is what 'allopatric' means). A river might form, or a mountain range rise, or a forest be divided by a desert, islands might be randomly populated, or, as we now know, a continent might split and move away. These things can happen rapidly, geologically speaking (to a geologist, a process that takes 20,000 years is considered 'sudden') or slowly (for geologists, basically any process that leaves strata, or layers, in the rocks). Each population will have to face slightly different conditions, and it will adapt to them in its own way, and time and chance will happen to them both. Each will go its own evolutionary way, so to speak.

For all the esteem that Darwin is held in by biologists, his view did not win the day. He thought species were simply well-marked and more permanent varieties. In the early twentieth century during the Mendelian revolution, it was the consensus that species were more than that, and that they were 'pure gene lines'. When the so-called Modern Synthesis of evolution and genetics was made in the 1940s, the views of Mayr and Dobzhansky that speciation was allopatric became the new orthodoxy.

Added to this, Dobzhansky was a firm proponent of the notion, developed by an American geneticist, Sewall Wright, in 1931 but with precursors back to Darwin, of *genetic drift*. Populations can be seen to be 'samples' of the entire genome pool of a species. Each population will differ slightly from its neighbours in the genes it has, and so if the flow of DNA between them is stopped by isolation, sequences that are rare in one may become, through simple random sampling, common in another. Species can thus be formed through random sampling 'error' if the differences that neutral genes cause prevent successful mating when they get back together; that is, when they become sympatric again. Note that this is not because there is selection for sexual or reproductive incompatibility, as selection acts on variants that are born and able to reproduce, not on those that cannot.

Recent work has brought Darwin's idea back into limited favour. It is now agreed that there is no single kind of speciation, and experiments and some historical cases have shown that sympatric speciation can occur. A fruit fly in the United States and Mexico, *Rhagoletis pomonella*, that lived on the hawthorn (or dogwood) tree shifted over to introduced apple trees in California during the nineteenth century. Since the introduction date of apples is known (not by Johnny 'Appleseed' Chapman; he did not get to California), it is known how long this took – and it appears that it took less than 50 years, though many of the genetic changes that were involved had been around as races in the flies for some time. The two varieties are now isolated, and do not mate, for a simple reason: the hawthorn and the apple bear fruit at different times of the year. Hybrids are neither adapted to the timings of one tree nor to the other, and are thus less fit than either variety, forcing them to adapt not only to their host trees, but to the preferred time of year each has for mating. Thus, they are forced apart.

It's arguable that *is* allopatry – as the host trees are, in effect, different territories, which is how the allopatrists (every term must have an -ism) explained it at first. But if that tack is taken, then there is effectively no difference between sympatry and allopatry, and it undercuts the notion of ranges. It might be better to call this allochrony, as the separation is seasonal. In any case, as explained so admirably by the Dutch invertebrate biologist, Menno Schilthuizen, in his book *Frogs, Flies and Dandelions*, sympatry is back on the menu of speciation modes.

However, this objection raises an important point. In population genetics, the field of such matters, the distinction between *patria* is not so much about place as it is about the migration of mates. Population genetics has a migration variable m in terms of the integration of the gene pools. But this is not even about the migration of organisms. In the end it is about their *alleles* (that is, variant sequences of genetic material in one spot on a chromosome) that move between populations successfully (and populations are not always, as I noted, in different places). The movement of genetic material, and its subsequent success, is what speciation is about, at least among sexually reproducing organisms.

So, what about the rest of the word salad? What does all that mean? What, for example, is *peripatric* speciation? Well, it was not long before it was realised that regions can also adjoin each other, like on either side of a narrow access such as a pass or unfrozen ground. If a population is outlying from the main population, speciation can occur when there is almost no gene flow. Hence Greek *peri-* (round or near to) and *patris*, peripheral populations. Mayr argued that peripheral populations have greater divergence and difference than central populations of a species range, based on his work on birds. So, a population at the extreme edge of the range will be more likely to evolve independently than one in the centre of the range.

What about *parapatric* speciation, then? Peri- and parapatry are sometimes used as synonyms, particularly with respect to speciation modes, but they do differ. Peripatric speciation is caused by being at the edge of the range and almost isolated geographically. But parapatric speciation is where the process of becoming genetically isolated (failing to share genetic material) causes the population to become geographically isolated. The first one has the geography leading the genetics, but the second has the genetics leading the geography.

Chromosomes

Stasipatric speciation is an intriguing one. An Australian entomologist, Michael J. D. White, specialising in grasshoppers, noted in the 1960s that there were races of a South Australian grasshopper (a race, to a biologist, is just a group found within a species) that had the same range, but had markedly different chromosome counts. Moreover, they didn't differ in their adaptations or food requirements. They just had double or triple or even more the number of chromosomes of their fellows, and more to the point, they could not interbreed with the single copy number grasshoppers. In one place (Greek: *stasis*), they formed new species.

In fact, the standout feature of stasipatric speciation is that a new species can arise in the midst of another. This often happens when entire sets of chromosomes duplicate but are retained in the sex cells rather than divided in half as normal, leading to individuals with a different chromosome number (called *ploidy*). The multiplication of chromosomes is called *polyploidy*. It happens that a good many organisms form new species by the multiplication of entire sets of chromosomes, especially ferns, corals and flowering plants, and many do this in the act of hybridising between related species. Stasipatric speciation is a major process of causing new species for a lot of organisms.

Varieties of ploidy include *polyploidy* (whole numbers of chromosome copies, usually written as $n = 38$, etc.), *haploidy* (one copy of all chromosomes rather than two; what sex cells are supposed to have), *diploidy* (two copies of all chromosomes; what sexual organisms are supposed to have after the egg is fertilised by the sperm), *alloploidy* (chromosomes from a different species) and *aneuploidy* (when there are too many or too few chromosomes for the normal form of that species; one example in humans is Trisomy 21, or Down's syndrome, which results from one too many chromosome 21s).

Rounding It Out

Which leaves us *dichopatry*. What on Earth is this? My Greek lexicon tells me that the prefix *dicho-* means to split asunder, or to have two adjacent things, as in dichotomy. This means in biology the kind of speciation where a species more or less evenly divides across its range, such as by a river or mountain range. The north side might evolve to become one species, and the south side the other, as seems to have happened in the chimpanzee/bonobo split across the Congo River one to two million years ago. The mechanisms are unclear if they aren't the same as allopatry, but I suppose it rounds out the set of alternatives.

All of this goes to show the complexity of speciation, and most importantly, that species do not gradually change as a whole from one species to another, and that within the research programme and theoretical tradition since Darwin, they never have. Being a species is more about maintaining genetic *homeostasis* (since we're doing the Greek etymology here, work it out: *homos* means 'same', and we have already encountered stasis; so, staying in the same place or state). In short, when a species has its own range, it's all about homeland security.

To understand all this, perhaps a metaphor might help. Imagine my favourite pizza, ham and pineapple (don't judge me). It represents the range of a whole species. Now move a slice out a bit but leave the cheese connected: that's peripatry. Separate it entirely, and that's allopatry. Now assume this pizza is from a dodgy pizzeria, and it comes with bacteria in the sauce. On the whole pizza, the bacteria interbreed (this is a metaphor, so the fact that most bacteria do not sexually reproduce can be ignored for philosophical clarity) and stay more or less the same. They have even adapted to eating the pineapple. The separated pieces, though, get only a sample of the genes, and so the bacteria there may change to prefer crusts. And if a bacterium happens to divide its genes but not split, so that there are bacteria with larger chromosome counts, then that is stasipatry. The moral of all this is, refrigerate your pizzas.

Other Kinds of Speciation

There are a few kinds of speciation that do not involve either adaptation to an environment or sexual reproduction. One is asexual speciation, which I will address next. The rest are in a grab bag category of 'other' (some advice: never tick the box marked Other, since it has to be dealt with manually or individually. This is as true in systematics as in bureaucracies). They all involve environmental factors, however.

Non-genetic speciation involves an existing population having variation, also called a norm of reaction, in which without any genetic change needed, individual organisms of the species can vary their body (their *phenotype*) to individually adapt to the conditions in which they develop, by way of what biologists call a switch mechanism. So, a plant may have a phenotype that adapts as it grows to dry weather and cold weather differently. Individual adaptation will not form a new species, of course, but if the norms of reaction vary in the population, more successful genetic variants will have an increased probability of passing through to the next generation. Over time, this, including any novel varieties that may happen to arise, will become the new average, or mode, for a distinct population. This does involve natural selection at the end stage, but the changes are caused by the ways in which genomes are expressed in the first instance. It is often referred to as individual adaptation, as opposed to populational or genetic adaptation.

Such phenotypic changes are called *phenotypic plasticity*, which basically means that the development of organisms can be very individually changeable. If the environment changes rapidly (geologically/evolutionary speaking) then those populations of organisms that are more able to change individually can respond in ways that leaves the population robust when things change. If genetic expression were rigid, then it would be fragile under environmental change, like a sheet of glass in a warped frame.

One other form of plasticity is behavioural. Not only the physical form can adapt individually, but organisms' behaviour can change their environmental conditions. There are forest elephants in central Africa (a third species, by the way, from the African and Indian elephants, called *Loxodonta cyclotis*) that have modified their environments so that they can get to the mineral and salt deposits deep in the forest to make up for a lack of trace elements and salt in their diets (forests tend to be salt- and mineral-poor). They have, as elephants tend to, made trails, which are maintained by continual migrations, and thus are passed down to their progeny, a form of inheritance of material conditions known as *niche construction*. As a result, at least in part, they have flourished in the forest and have undergone selection for smaller size and straighter tusks to move around more freely.

The most obvious kind of niche construction to us humans is cultural. Human beings are excellent niche constructors. From the use of fire to change plant ecotypes, the forms of the general plant life in an ecosystem, to improve hunting conditions (probably even before *H. sapiens*), to making our own trails and the information transmission of these to later generations to improve the efficacy of foraging (yes, and hunting) seasonal foods, humans construct their worlds. Such constructed environments are our native conditions, and in my view, this applies to post-agricultural conditions like cities as well. We are now, and always have been, in our state of nature.

Niche construction is a version of 'Baldwin effect', named for psychologist and philosopher James Mark Baldwin at the end of the nineteenth century. Baldwin proposed that animals can change their behaviours, creating conditions that natural selection would need to 'catch up' with, mutations permitting. He distinguished 'habit', 'assimilation' and 'accommodation'. The first two depend upon existing genetic capacities; the latter, selection for the best able to do the assimilation. Whether or not this leads to new species is contestable. I am not aware of any case that has been well-attested, although some evolutionary biologists are working on the topic.

Another 'other' speciation type is parasite-driven. We mentioned the apple and hawthorn flies above as adaptation to new habitats, but from the trees' perspective, these are parasites. If they changed the host's ability to interbreed with unparasitised trees, that would be a case of parasitic speciation. And it turns out there is just such a widespread form of this speciation, among wasps, bees and flies. A single-celled parasite named *Wolbachia* that inhabits the cells of the host (I mean, literally invades and lives inside the cells) can cause infected hosts to be fertile only with other infected hosts by infecting the sex cells (the *gametes*) of insects, making them unable to successfully fertilise or be fertilised by uninfected gametes. Such infections sometimes do offer some benefits to the hosts in terms of resistance to other infections, and so over time they may become commensualists (that is, mutually beneficial with their hosts), although pathogens and hosts need not evolve that way.

A final point before we move on: people often talk of organisms behaving in a manner for the 'good of the species'. Vero C. Wynne-Edwards, the British ornithologist, argued, just as with reproductive isolation, that organisms that, for example, lay fewer eggs in hard times cannot be the result of natural selection acting only on individuals. Therefore, he said, it is the result of selection on larger groups, perhaps even species. This view, called group selection, was solidly attacked by more 'orthodox' evolutionary biologists, leading to, for example, Richard Dawkins' view that evolution was all about genes, not species, in his 1976 book *The Selfish Gene*. The reason that the orthodox object to group selection seems to be that if there are forces causing group adaptations, they can override the forces acting directly on genes and individual organisms, thus overthrowing not only Darwinian evolution but the very idea of causes working from part to whole. The debate rages rather vapidly lately, although E. O. Wilson was still a group selection proponent of sorts at his passing.

There are several other forms of 'other' types of speciation, but the last one I will discuss is the formation of asexual species.

What's Sex Got to Do with It?

Once upon a time, species were thought to be mostly sexual, and few if any weren't. This long dark age ended around 1960, although specialists in the relevant fields knew well before that there were asexual organisms. Increasingly, single-celled organisms, which used to be called Monera before the revolution in bacterial classification from the 1960s onwards, were seen to share genes infrequently and only partially. First there was a distinction between prokaryotes and eukaryotes (organisms without membranes around the nucleus of their cells, and organisms with, respectively), and then prokaryotes were divided into three kingdoms: Archaebacteria, Eubacteria and Protista (the last is a eukaryote group of single-celled organisms – a bit of a trashcan category). At last report, the Linnaean classification of life includes two empires (prokaryotes and eukaryotes) and seven kingdoms (with the capitals, they are the formal names of those kingdoms; without, the words are general terms):

Prokaryota	Bacteria
	Archaea
Eukaryota	Protozoa
	Chromista (algae and diatoms)
	Plantae
	Fungi
	Animalia

Note that these are considered conventional – not natural – groups by many biologists. Also, they change rapidly, so this may already be out of date. Viruses aren't shown here because most do not think they are alive.

Now, while Bacteria and Archaea generally lack sexes, they do exchange genetic material in one of three ways: *transduction*, *conjugation* and *transformation*. In transduction, one bacterium receives genes from another via viral insertion. Some viruses can take up segments of genes from one host whose genetic machinery they are hijacking, and inadvertently introduce those segments into another, not even closely related, cell. Even in animals this is possible (in animals this process is called *endogenous retroviral insertion*) and as much as 42% of the general mammalian genome has been identified as the result of ancient and modern insertions into the gamete lineage of a species and its predecessors; this includes humans. A virus, like any genetic molecule, has a sequence either of RNA or DNA depending on the nature of the virus (endogenous retroviruses are mostly DNA-based viruses, but most other viruses are RNA-based). Once in the relatively small genomes of bacteria and archaea, the inserted sequences can provide raw material for future genetic novelty – including antibiotic resistance (see below).

Conjugation is a bacterial process whereby one cell inserts a tubular structure known as a *pilus* into the contents of another cell's membrane and transfers genetic material in the form of a small circular structure of DNA called a plasmid to the recipient cell. This happens both within a species and between species, and a bacterial plasmid was once observed inserted into a hamster cell, so evolutionary distance appears to be no barrier at all. Conjugation is a reason why many bacteria can evolve antibiotic resistance: if it evolves in one bacterial lineage, it can rapidly spread to other lineages if the antibiotics are overused or improperly applied.

And finally, transformation is the process of taking up genetic material that exists in the surrounding medium. Single-celled organisms can release genetic molecules either by way of cleaning up during environmental stress, or by dying messily so that the contents of the cell spill out to be hoovered up by other cells.

This sharing of genes, however, does not much help the definition of microbial species (I'm using 'microbial' to cover both bacteria and archaea, and perhaps the relevant eukaryotic microbes as well). They are not as effective at maintaining clusters as sex is, and so they don't really explain the relative stability of microbial taxa. What does? So far as I can tell, there are several options offered by biologists: some genes are core genes that do the basic housekeeping of the metabolism of the microbe, and they are not spread to other microbes with a different set of housekeeping genetic materials, and so they keep the lineage distinct; or the lineages are literally tracking the adaptive zones or niches; or it's just a chance artefact of the fact that biology happens to be clumpy not smooth, and human observation tends to be short-lived. I suspect all three are right sometimes, but my money is mostly on the latter two – adaptive tracking and accident.

While we're on the subject of viruses, the question of what clumps they come in is even more unclear than microbes. Biologists will classify clumps of living (and in the case of viruses, adjacent to living) things no matter what, and so it seems that they will find something akin to a species everywhere. But viruses not only do not have sex, but they also do not even have cells. Still, they are classified, and within their 'species' virologists note 'strains' or 'variants'.

The definition of a viral species obviously doesn't involve genetic exchange or isolation, although 'superinfected' host cells (cells in the host organism that are infected by multiple strains) can swap viral genes when transcribing and replicating them. At first, viral species were defined as a cluster of strains, and later as a class of viruses that shares an ecological niche. From a viral point of view, a host organism is its environment, and what works for the virus is what makes a niche for it there. Whether or not that hurts or kills the host is of no 'concern' to a virus unless this reduces its transmission and thus reduces its fitness.

As a side note, it is for this reason that preventing infections usually causes a loss of virulence and reduction of host damage with pathogens, viral or not. If infection of the host population is difficult then there will be selection against those strains that rapidly exploit their hosts, and this will allow the more 'benign' (or at least less malign) strains to take over. Over time, prevention causes lowered virulence. If you were to allow an epidemic to rage unchecked by hygiene or vaccines, then the 'vicious' strains would keep outcompeting each other until they became a major threat to the hosts. Nobody wants that, right?

These days a viral species is considered to require both a single point of origin (it should be monophyletic) and to be distinguishable from other such lineages, to get a species name. This represents a step backwards in terms of theoretical completeness but may allow more species to be categorised using standard molecular genetic techniques.

But another concept offers help here. Manfred Eigen was a physicist who turned to biology, and he made useful contributions; one of them is the notion of a *quasispecies*. This word, which comes from the Latin for 'as if' (*quasi* + *species*, is something of a genetic abstraction, and it is a bit complex, but bear with me, because I think it has a real payoff. We all drew three-dimensional spaces in school. Suppose we made the sequence of a virus a coordinate in a graph space. Of course, even viruses have a number of 'genes' (with viruses, the actual work done by their genetic material is done by the infected cell, making it difficult to determine what is a gene and what is not in a DNA or RNA virus), but they seem to range from under ten to over a few hundred. If you make a graph 'space' with an axis for each 'gene', then strains would be points in that space. If you locate all the strains in this pace, you will find clusters, and an average (modal) 'centre' of the space. This is called the 'wildtype' genome, although there may be no actual strain that has it. And the clusters themselves are the quasispecies.

So, what's the payoff? It is this. Something like quasispecies underlies every definition of what species are, either in general or in a particular group of organisms. Species are clusters in gene-space at the most basic level. Or in phene-space (a graph of phenotypic traits), although that is less constrained by the observations, as the traits used can be chosen in a number of disparate and more or less informative ways, as the numerical taxonomy ('phenetics', on which more later) of the 1960s found out.

So that is all I have to say about microbial species, but what about multicellular organisms, or macrobes? How common is asexuality for fungi, animals and plants? Fungi can often reproduce sexually or asexually or both. Plants often reproduce vegetatively, which means that the budded plants are clones of the original stand. Such species are called *apomictic* in plants. Apomicts often are the result of cross-species hybrids, as they are 'gamete broadcasters', which means that rather than one-on-one mating, they rely on dispersal mechanisms such as wind, currents and pollination. Others are polyploids, an example of which is the Cavendish banana. Corals, which rely on the prevailing currents at the time they release their gametes ('sporulation'), can hybridise to form novel species when currents change. One noted coral specialist, John Veron, told me that the prevailing currents when spores are released are the most important cause of coral species.

OK, but animals must be sexual, right? Well... We can start with protists, many of which are asexual. But actual animals, sometimes grandiosely called metazoans? Yes. In fact, asexual animals, although rare, are common enough to give a name to: *parthenogens*. The term means 'virgin birth', only without the religious connotations. And they are very often also caused by cross-species hybridisation. The case study below discusses the asexual species of whiptail lizards (genus *Aspidoscelis*, see below), but there are parthenogenic snakes, birds, sharks, fishes, rotifers, nematodes, insects, scorpions and so forth.

Evolutionarily speaking, asexuality is a dead end, or at least it is supposed to be. This is because useful mutations that might help with an environmental challenge are unlikely to occur on the same lineage, whereas sexual organisms can share theirs. Therefore, many asexual species are low in genetic variance, and are vulnerable to, say, a disease. I mentioned the Cavendish banana above – as they are all clones of each other, a fungal infection to which that strain succumbs is busy destroying the world's crop.

But asexual rotifers (microscopic filter feeders that live in water) called bdelloids have been around for millions of years, presenting an anomaly. Sexual organisms can share advantageous mutations in their genomes (or at least share mutations that later turn out to give advantages, say in new climates). However, purely asexual organisms cannot share genes and so the likelihood that an advantageous mutation will arise in a lineage that already has other advantageous genes is much lower. Some work suggests that despite asexual reproduction, some of these lineages can alternate to sexual reproduction occasionally.

But with parthenogenic lineages like whiptails, that is very unlikely, although as we shall see below, even then the case is untidier than we might expect. In the end, as the Bible says, time and circumstance happen to all species (*Ecc.* 9:11), and we cannot predict how it will go.

The Case of the Asexual Hybrid Lizards

Whiptail lizards, or racerunners, are a group of lizards that are widespread throughout the southern United States down to Brazil. They occur within the family Teiidae, and in two genera, *Cnemidophorus* (South American, 46 species) and *Aspidoscelis* (North American, 19 species), and some 30% of these species are parthenogenic. *Aspidoscelis* was separated from *Cnemidophorus* in 2002.

In some species of asexuals, such as *A. neomexicanus*, hormonal expression changes the females to behave like males, and they engage in pseudocopulation which in the female-like female (they are all females) triggers egg laying, but without sperm. These eggs are fertile and produce clones of the mother.

The evolution of these asexual lizards, the largest group of vertebrate asexuals known, is due to both polyploidy and hybridisation (Figure 3.4). In the

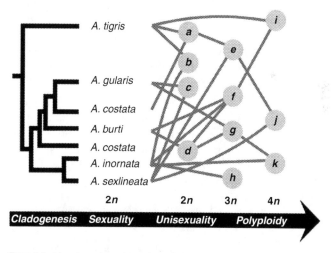

Figure 3.4   How to make unisexual species.
Sexual and unisexual species forming by hybridisation followed by polyploidy in the whiptail genus *Aspidoscelis*. The species are: *a*, *A. tesselata* complex;
b, *A. neomexicana*; *c*, *A. laredoensis* complex; *d*, intermediate ancestor;
e, *A. neotesselata* complex; *f*, *A. flagellicauda* and *A. sonorae* complexes; *g*, *A. exsanguis*;
h, *A. opatae*, *A. uniparens* and *A. velox* complexes; *i*, 4*n* by *A. sonorae* (female)
× *A. tigris* (male); *j*, 4*n* by *A. neotesselata* (female) × *A. sexlineata* (male); and *k*,
4*n* by *A. exsanguis* (female) × *A. inornata* (male).

figure, the mid-grey lines show which sexual species hybridise to form either single-sex 'species' (asexuality) or species with many copies of chromosomes. The $2n$, $3n$ and $4n$ labels indicate the number of copies of the chromosomes each type has. To say this is complex is a bit of an understatement.

Now, if these asexual species are not species at all, this means that in the clade of *Aspidoscelis* species around 30% of the taxa in the group are not species, an odd conclusion to say the least. Some groups would not be species for an arbitrary reason of the choice of definition. But if they *are* species, then a singular definition of species and speciation cannot be found in just that group. And they are far from the only group of animals in which this happens.

So, it seems such cases require both the 'normal' allopatric cladogenic speciation mode and the stasipatric mode. The difference with hybrid forms and the usual 'Darwinian' speciation is that hybridisation is *reticulate* (that is, it forms a web, and not a tree). This is often referred to as horizontal (or lateral) genetic transfer (HGT or LGT), since it moves across branches of the tree rather than along them. LGT usually involves a low rate of gene sharing, via introgression, but hybrid lineages can include a much greater proportion, up to 50% from each contributor.

Reticulation (horizontal connections; its antonym is *articulation*) means that it is very hard to establish the evolutionary history of a complex. However, the genes themselves can have articulate histories, so when genes can be sequenced, it is a lot easier to establish who contributed what. This is to say that gene sequences are mostly related through a tree pattern, while organisms and even species are related by web, or network, patterns. Lateral transfer, however, reintroduces the issue of reticulation to classifications, of species or anything else.

So that is a summary of what *makes* species according to modern biology. There may be novel processes in the future: genetic modification by humans, perhaps? However, unless humans have some clairvoyant skills, and can predict what will happen over the long term, which we cannot, I think we will act like another environmental influence, and the speciation types will remain pretty well unchanged.

4 A Short History of Species and Kinds

Textbook histories are how most scientists learn about the past of the ideas and disciplines they employ, and any textbook will tell you that the idea of species goes back to the classical era if not earlier. In a way this is true, but textbook histories are written by scientists, not historians, and they often repeat untested or false ideas for reasons other than knowing the past. Often, history is something to be used as a way of establishing the in-groups and out-groups of science; in other words, history can be used as a weapon in the sciences. So, some critical revision is required.

Plato's theory of Forms uses a closely related term 'idea' as well as *eidos* to denote 'forms', which are eternal and beyond the physical. Plato, as with philosophers since who are interested in kinds of things, used biological illustrations, such as 'horse', 'human' and 'dog', but he did not think actual horses, humans and dogs were species (or members of a class of things) because none of them, not even Socrates himself, were perfect examples of their forms.

Aristotle, who seems to have thought that forms were always 'real' (in the sense that they are kinds in physical form, or instantiated) also used *eidos* for forms, but he did not use that term for kinds of living things in his 'natural history' books, the *Historia Animalium*.[1] Instead he said (in Greek, of course) they were 'of the same tribe' (homophūla) or 'same-born' (homogenēsin). His

---

[1] That is not to say he never used these words when referring to what we would call a species, but he means them as 'groups' or 'kinds' in general. See the beginning of his treatise *On the Parts of Animals*, around 645b.

student Theophrastus, who is often called the founder of scientific botany (which is in my opinion a bit silly – no discipline has a single founder), likewise used various terms other than *genos* or *eidos* for living kinds.

The term *species* in natural history (what we now call biology and the Earth sciences) had no special significance until the seventeenth century. Species, like genus, was just an ordinary word in Latin for a term that meant a kind. In 1686, when preparing a complete study of plants, the Reverend John Ray wrote that he needed some way to identify kinds (species – he wrote in Latin) of plants, and he came up with the definition 'distinguishing features that perpetuate themselves in propagation from seed'. He extended this also to animals, and thus was a universal natural concept of species in biology born.

The words species and genus were not only logical terms but were also ordinary language terms (to a Latin speaker) that meant 'kinds' or 'sorts', and they were used interchangeably (as also were their Greek equivalents *eidos* and *genos*) for all kinds (see?) of things, living and social and astronomical and geographical. Calling something a kind or sort of thing had no metaphysical connotations, such as things being idealised forms or having essences, then or now. So, when educated Latin-speakers in the seventeenth century started to settle on terms for kinds of plants or beasts or birds, they chose genus for the enclosing group and species for the enclosed group, or their vernacular equivalents (Table 4.1).

In Table 4.1, I've also included phrases for living kinds and organisms for comparison. The term *organism*, and the cognate terms in other languages, did not come into wide use to mean living beings until the early nineteenth century, first in French, and then taken over by English and German, and so on. Instead, people spoke of 'organised beings' – literally beings with organs. Moreover, Greek terms used in antiquity (in this case by Aristotle) translate as 'same origin' or 'same tribe'. In older English and French, types of organisms were often 'families' or similar ordinary terms. Words and their etymology alone do not prove anything (look up the etymology of 'let' sometime), but it is quite clear that nobody was doing any metaphysics just because they used terms like *species* or *genus*. These were just ordinary terms for kinds without any technical meaning outside of logic.

Prior to Ray's definition, botanists, and to a degree zoologists, interchangeably used terms like *genus*, *species*, family, or simply the local words for beasts and

| Language | Term used in English | | | |
	Species	Genus	Living kinds	Organism
Greek	Eidos	Genos	Homogenē, homophūle	zoō, fytó
Latin	Species	Genus	Genus	animālis, planta
French	Espèce	Genre	Genre vivant	animal, plante (later: organism)
German	Art	Gattung	lebende Art	Organismus
Spanish	Especies	Genero	Tipo de vida	organismo
Older English	Species, form, kind	Tribe, family	Creature, being	Organised being

Table 4.1   Terms in biology from the Greeks to Old English

plants. The technical kind in logic was *species*, and adopting the term as a 'term of art' (see Chapter 5) in natural history also licensed the use of *genus* for more inclusive groups. The seventeenth-century Swiss botanist Caspar Bauhin had used the *genus–species* pairing, but he used it inconsistently and he did not intend it to be a formal binomial name. Most likely he did so because educated folk of the time knew of Aristotle's scheme for predicates (the meanings of words), where a general term included special terms.

The pairing was adopted by Carol (or Carl, or Karl – spellings were optional at the time) Linnaeus (or Linné) in his systematic arrangements that came to be known as the Linnaean Natural System. Ironically, although Linnaeus made *species* the 'unit' of his system, he never really defined it, taking instead the tradition started by Ray as his foundation. He did take one other thing from Ray – the notion that species were today as God made them at Creation.

During the Middle Ages and the classical era of Greece and Rome, natural species were somewhat fluid. As the Reformation in the sixteenth century occurred, the tradition of taking the Bible's account in Genesis literally was getting underway, and this led to the problem of reconciling the text's phrasing *after their kind* in chapters 1–2, in the *Vulgate* translated either as *species* or as *genus*, as applied to the legend of Noah's Ark, with the burgeoning awareness of the many new kinds of plants and animals from outside Europe. One solution that various scholars (such as Athanasius Kircher, a Jesuit priest) adopted was that the Ark included the 'created kinds' or 'original kinds', which then massively hybridised, adapted geographically, or even spontaneously generated (worms and insects, which did not have 'breath', could arise this way) to form the multiplicity of observed species today. As an aside, many modern creationists also propose not an anti-evolutionary but a hyperevolutionary process post-Ark for the same reasons.

But within seven years of Linnaeus publishing his first edition of the *Systema Naturae* in 1735, a surprisingly modern evolutionary view was anonymously published by the later famous French physicist Pierre Maupertuis (the formulator of the least energy principle among other achievements) under the pseudonym 'Dr Baumann' ('builder'). In his book *Vénus Physique* (*The Physical Venus*) he suggested that heredity was caused by unchanging particles, and, based on his study of polydactyly (more than five fingers) in French families, he came very close to Mendel's ratio (see Kostas Kampourakis' book, *Understanding Genes*, in this series). He lacked only a theory of natural selection to be broadly up to date with the modern evolutionary synthesis (see below). Maupertuis' view of species as able to change did not generate any scientific programme, but at least it indicates that despite the popular view that before Darwin species were always thought to be fixed (static), this is at best a simplification (which Darwin himself contributed to in his 'Historical Review' preface in later editions of the *Origin*).

And there were several 'evolutionists' between Maupertuis and Darwin: Geoffroy Saint-Hilaire, Lamarck, Erasmus Darwin, Robert Grant, Robert Chambers and Patrick Matthew. Jean Baptiste Lamarck, professor of zoology at the National Museum of Natural History in Paris, is the most famous evolutionist before Darwin, but he had a perfectly standard definition for species:

> ... in botany as in zoology, a species is necessarily constituted of the aggregation of similar individuals which perpetuate themselves, the same, by reproduction. (Lamarck, quoted in Britton, 'Taxonomic aspect')

Where he differed from his contemporaries was that he simply did not think that species were natural objects; that is, that they did not exist. Instead, he held that each present 'species' was the result of a spontaneously generated 'monad' that, through the acquisition of individual adaptations, changed as a whole over time to become more complex, and that human beings were the most complex. He thought that the genealogical lineages of these monads changed all at once, driven by an inner force or 'ethereal fire'. Incidentally, Lamarck's notion of the inheritance in progeny of traits acquired during their parents' lifetimes was not original to him, nor was it particularly controversial until a century later.

Étienne Geoffroy Saint-Hilaire (usually referred to as Geoffroy) was a naturalist at the National Museum of Natural History in Paris who worked on embryology, classification, and ideal types among animals. He disputed with Baron Georges Cuvier, who was professor of anatomy at the Museum, over whether species could transform over time, which he came to believe they could after Lamarck's views had been popularised. Cuvier had been the strongest opponent of transformism, as evolution was then called, in part because he thought that Lamarck's view was contrary to the evidence (Egyptian mummies of cats were the same as modern cats). Cuvier gave the most influential definition of species for the nineteenth century across Europe, in 1812:

> My research assumes the definition of species which serves as the basic use made of the term, understanding that the word species means *the individuals who descend from one another or from common parents and those who resemble them as much as they resemble each other.*
> ('Discours', p. 125)

Cuvier's later *Éloge* (obituary) for Lamarck was basically a hit piece against a dead opponent, who was therefore unable to defend himself, after Lamarck had died in 1829 blind and in poverty. His daughter eventually had a statue raised to him in the grounds of the Museum. Cuvier's view influenced many, including Darwin's friend Charles Lyell, while Geoffroy influenced Darwin's opponent, Richard Owen, a vociferous anti-evolutionist. Such are the ironies of history.

Robert Grant was one of Charles Darwin's lecturers in medicine at Edinburgh University. He was an admirer of Lamarck's and told Darwin so. Another evolutionist was Erasmus Darwin, Charles' grandfather and a doctor and a widely respected man of letters. Neither had a great influence on subsequent science. Robert Chambers, however, did. A publisher and author of historical texts, he anonymously published *The Vestiges of the Natural History of Creation* (1844), which caused a great sensation. Although he had nothing much to say about what species were, he at least asserted that they were not fixed in a particular form.

Darwin and His Successors

There is a tendency among biologists to assign to Darwin all that is good and modern in our understanding of nature, especially regarding species. As a historian of ideas, this is something I dispute. Yes, Darwin effected the greatest mindshift in modern thought about living things, but he is not responsible for every shift of thinking.

The continuity of human beings with the natural world, the competitive aspect of nature, the treelike aspects of diversity are all pre-Darwin. Some of this is due to Alexander von Humboldt, the polymath Prussian explorer, who deeply influenced Darwin. Some of it is due to the French botanists the Candolles, father Augustin Pyramus and son Alphonse, who identified the variation within species and the war of nature, and so on. The first tree-metaphor for living kinds was proposed by Peter Simon Pallas in 1766. There are many sources in science for any idea.

Darwin's genius was synthetic rather than analytic or innovative. He brought together strands of thought that were in general pretty correct and created an overall view, 'this view of life' as he said in the *Origin*'s last pages. It was this unified vision that so influenced those who followed after him, especially Ernst Haeckel, who coined the term 'ecology', and numerous twentieth-century ecologists such as Britons Arthur Tansley and Charles Elton, the Dane Eugen Warming, and the American Eugene Odum, all of whom influenced science, popular opinion and policy.

Taking care not to assign to Darwin views he either did not originate, or did not hold, such as the Romantic view of Nature that Humboldt had promoted, or the notion of the continuity of humanity with the rest of the living world (much older than even modern science), we can appreciate how much he actually did achieve. In particular, he made it obvious that life was old, and changed over time, in ways that created the diversity around us, and implicit in his work was the thought that once a species goes extinct, it is gone forever. That is possibly the most significant thing he ever said.

In late antiquity and the early Middle Ages (around 300 ce to 800 ce), Latin was the *lingua franca* (a term from when French itself was the Latin of its day, much later on; today the *lingua franca* is English, but Mandarin looks like a coming contender) of educated people and merchants in western Europe. And later, from the ninth century to the thirteenth century, there was a general movement to translate books into Latin, out of Arabic, Greek, Hebrew and other languages, so the West could read them (the eastern part of Europe spoke Greek or similar tongues). Aristotle's works were translated. The Bible (the *Vulgate*, which means 'the common tongue') had been translated by St Jerome in the late fourth century, and in it and the other works, 'kind' terms in the original texts generally got interchangeably translated as *genus* and *species*, especially in logic and theology. So, by the sixteenth century, these two words were the go-to terms for kinds.

But these were not yet technical terms in natural history. In fact, if they were technical terms in any field, it was in logic, where a general kind was a *genus* and a special kind was a *species*. Moreover, a special-kind species could also be a genus, called a *sub-altern genus* (literally, 'kind below everything'). When the process of logical division got down to the point that any further division didn't give kinds, but only individual things, you had the lowest species, or in Latin, *infimae species*. Think of them as sets that have subsets, which in turn can be sets that have subsets, only without all the Venn diagrams (actually, they were very often shown as tree diagrams or indented lists).

All of this is prelude to the history of *living* species. Nearly all the claims that textbooks make about species being universally thought to be fixed (that is, unable to change beyond a certain amount) and isolated (being unable to interbreed with other species) are historically false, rather like the claim that during the Middle Ages religion interfered with studying the natural world. It's the natural history equivalent of the false belief that the medievals thought the world was flat. Although in the eighteenth and nineteenth centuries it was not uncommon to find people who did believe in fixity, neither was it uncommon to find naturalists thinking that species were not fixed.

It's often said that the Middle Ages were a time of darkness and ignorance, but this is only true, if at all, in one relatively small part of the world and only for about 400 years after the Roman Empire collapsed in the western half of Europe. Intellectual life flourished over the centuries from Charlemagne in the ninth century, to the period preceding the unfortunate pandemic of the Black Death in 1347 which killed about one-third of the West's population, curtailing, among other things, an extensive interest in the natural world and many other academic pursuits such as logic, on account of there being other priorities and many of the authorities in these fields being, frankly, dead.

Around this time, the Bible started to be treated less metaphorically (or *allegorically*, as the church called this mode of interpreting the Bible) and more literally. By the fifteenth century, this meant that interpreters were trying to find ways to make the biblical narratives align with the observed world. One obvious and glaring difficulty was how many kinds of animal could fit into Noah's Ark. After all, the Bible gives its dimensions (*Genesis* 6:14–16) and some information about its management for a year on the water, and since the Flood covered the mountain tops, all land animals that now exist must have been on it. However, the logistics increasingly failed to add up. There were far more animals than could have fitted on even the biggest interpretation of the Ark (there was some debate about how long a cubit, the measurement used in the Bible, really was), and so there needed to be some theological adjustments. Things that did not have 'breath' (*ruach* in Hebrew, *pneuma* in Greek, and *spiritus* in Latin) were not a problem as they were believed to arise through spontaneous generation from mud; and nor were fishes, which supposedly could live in any water (even the freshwater fishes were supposed to be able to live in the diluted salt water of the Flood). Plants were left to their own devices, sometimes because the theologians agreed with St Augustine that the Earth itself has plants *in potentia*. But animals in general were difficult to accommodate, especially mammals and birds. The solution, offered by, among others, sixteenth-century French mathematician Johannes Buteo (or in French, Jean Borell), was to take a passage out of Aristotle's work and ascribe the broader variety of kinds (either genera or species) to cross-kind matings. In short, hybridisation.

By the seventeenth century, studies in botany, and to a lesser extent zoology, were well underway, and the number of kinds expanded some more. John Ray had decided to give a list of the parts and characters of plants in a three-volume set entitled *A History* [meaning an Investigation] *of Plants*, beginning in 1686. Before this text, the term *species* just meant some sort or kind of organism, although most naturalists had settled on it as a shared term. When arguments began about whether or not there were one or more species for this or that group, it became important to know what was meant by the term *species* in natural history. That is, a distinctly biological concept of species was needed, and Ray gave it here:

> So that the number of plants can be gone into and the division of these same plants set out, we must look for some signs or indications of their specific distinction (as they call it). But although I have searched long and hard nothing more definite occurs than **distinct propagation from seed**. Therefore whatever differences arise from a seed of a particular kind of plant either in an individual or in a species, they are accidental and not specific. For they do not propagate their species again from seed ... So, equally in plants, there is no more certain indication of a sameness of species than to be born from the seed of the same plant either specifically or individually. For those which differ in species keep their own species for ever, and one does not arise from the seed of the other and vice versa. (Ray, quoted in Lazenby, *Historia Plantarum*; my bold)

Note that this is not a 'biological' species concept – there is nothing here about interfertility; it is a *generative* conception of species – living things are the same species if they generate the same 'forms' reliably. This gave naturalists a clearer idea of what it was to be a species of living kinds, for both plants and animals. But this, while useful to the systematist (one who organises a taxonomy into a system), did not address the *causes* of species. They were assumed to have been created by God in the beginning (Ray was, after all, a minister of religion), and this was expressed overtly by Carl Linnaeus around 40 years later. Also, Ray is the first person to clearly express the fixity (the static view) of species, using the Aristotelian distinction of essences and accidents (see Chapter 5). This means that the idea that species were once created as they are now was not the default view amongst naturalists until much later.

Ray's ideas were in some manner a reaction to the Noah's Ark logistics of Bishop John Wilkins (no close relation to your humble author), who was

Oliver Cromwell's brother-in-law. Wilkins was an exponent of the idea of a Universal Language and the last Ark logistics exponent (until very recently, when creationists revived that discussion in the early twentieth century), and he employed a young Ray to squeeze all species into his *a priori* scheme. Ray was stung by professional criticism by a leading botanist (Robert Morison at Oxford) and took a more empirical approach from then on, leading to the first regional *Flora* (a book that gives all the plant species of a region). So, in one way, we only have a concept of *living species* due to debates over Noah's Ark.

The Rise of Systematic Biology

After Ray, botany got serious about its classifications. To a lesser extent, zoology had got serious about it too, following on from the *Historiae Animalium* (*Investigations into Animals*) of Conrad Gesner, in the late sixteenth century. Gesner also began a similar work on plants but died before he could finish it. Both works had numerous illustrations, including the famous rhinoceros etching by Albrecht Dürer (see Figure 4.1). Gesner's watercolours of plants are so realistic, one could take them into the field and instantly identify their targets. He and his collaborators presented the plant throughout its life-cycle, unlike later botanical illustration which focused on stereotypical mature specimens.

It was in botany that classifying really took off in the seventeenth and eighteenth centuries. Gesner and Ray both listed organisms in a random fashion, although Gesner did arrange animals into the folk categories of fishes, birds, and egg-laying and live-bearing four-legged beasts. However, there was no systematic arrangement of species – just a list alphabetically arranged by the Latin name. Botanists needed to find a way to make sense of all these species. Thus was born *systematic botany*, and later *systematic biology*. The term 'biology', like the term 'scientist', is a nineteenth-century invention, so at this time (seventeenth to eighteenth centuries) the correct term is 'natural history' for Earth-based sciences ('natural philosophy' covered the physical sciences and astronomy).

Linnaeus, a Swedish medical doctor turned botanist, organised species in terms of relatedness based on shared traits, and this, first published under the title *Systema Naturae* in 1735, became known as the 'natural system' in the

Figure 4.1   The imaginary rhino.
The Rhinoceros, by Albrecht Dürer, 1515. Dürer drew this from sketches and descriptions of a dead rhinoceros he did not see himself. Nevertheless, it is accurate enough to identify it as an Indian rhinoceros.

early nineteenth century. This was especially true in Britain, where the Linnean Society was formed by Sir James Edward Smith, a botanist himself, to store his specimens and to enforce his classifications. Since Linnaeus' scheme included *genus* and *species* in its ranks, species became the default unit of classifying living things thereafter, and so the question of what a species *was* became a critical issue, leading a youngish Charles Darwin to become part of a committee (the Strickland Committee, named after Hugh Strickland, who chaired it) which formalised systematics. This later evolved into the International Code of Zoological Nomenclature (ICZN) and the International Code of Botanical Nomenclature (ICBN), and others followed for bacteria, fungi, viruses and so on (see Chapter 6 for the current state of systematics).

Initially the solution to what should be included as a species in this code was practical. In order to prevent every plant or animal breeder from naming any variety as a species (which often happened to increase sales, especially among the gardening-crazy British of the nineteenth century), the 'right' to name a species was restricted to a museum-based taxonomist, no matter who collected the specimens. This was later broadened to a 'competent systematist' whether or not they worked in a museum.

At the time Darwin wrote (around 1844 to 1880), when his publisher John Murray put the word *species* at the forefront in his book title (1859), botanists and zoologists were arguing at length about whether some organisms were species or merely varieties. Darwin's own solution was that something was a species if it was a variety that became less temporary, although he never quantified the duration. Contrary to a good many scientists and historians since, Darwin did not deny that species existed. He did deny they were all of the same type of thing.

Darwin thought that species were formed by natural selection (as we discussed in Chapter 3), but he got into a long debate with the German naturalist Moritz Wagner who argued that species formed by geographical isolation (a view Darwin had entertained in the 1840s but dropped). Darwin grudgingly gave some ground, but remained convinced that natural selection was the main, if not the only, way new species were formed. The subsequent debate raged, if one can call polite notes in scientific journals 'raging', for several decades, and was revived in the 1940s by Ernst Mayr, the German ornithologist who migrated to the United States as we noted in Chapter 2.

Interestingly, at the time of the development of modern Mendelian genetics, from 1900 onwards, the notion that Darwin thought species were unreal arose, in part because he didn't give a solid definition of them, and in part because the Mendelians tried to do just that. Wilhelm Johannsen, a Danish geneticist who was one of the 'founders' of genetics (though there really are no such things as unique founders in science), proposed that the real units of biology were what he called 'pure lines' – discrete genetic kinds that formed by mutations. Thus, Johannsen is responsible for the 'biology' underlying the X-Men. Later, the phrase 'hopeful monsters' would be suggested by Richard Goldschmidt, a German biologist in the 1940s. This is sometimes known as *macromutationism* or *saltationism* (from the Latin *saltus*, meaning a leap).

The problem here was that Mendelian genes seemed to breed true, and so there could not be the continuous variation needed for natural selection to favour some varieties over others. It took a number of mathematically inclined geneticists, notably an American, William Castle, and a Briton, Ronald Aylmer Fisher, to show that many genes in concert could cause almost continuous variation on which selection could act. Nevertheless, the Species Problem, as it was now called, had been set up, and from that context we now have a full-fledged debate.

In 1935, the Russian geneticist Theodosius Dobzhansky ('Dobie' to his friends) who migrated to the United States on a scholarship in 1927, published in a philosophy journal a paper entitled 'A critique of the species concept in biology'. He asked: *Is, then, the species a part of the 'order of nature', or a part of the order-loving mind?* That single question is still being debated today. Dobzhansky answered it by saying that isolating mechanisms that kept species apart when in geographical contact was what made a species. He then defined a species as:

> that stage of the evolutionary process at which the once actually or potentially interbreeding array of forms becomes segregated in two or more separate arrays which are physiologically incapable of interbreeding.

Shortly afterwards, in 1940, Ernst Mayr wrote a paper, 'Speciation phenomena in birds', in which he defined species as populations that are replaceable by each other when in contact and interbreed:

> A species consists of a group of populations which replace each other geographically or ecologically and of which the neighboring ones inter-grade or hybridize wherever they are in contact or which are potentially capable of doing so (with one or more of the populations) in those cases there contact is prevented by geographical or ecological barriers.

This later became known as the *biological species concept*, which we have already met. Even though Mayr wrote this of birds, he quickly generalised it to all living things. Dobzhansky's version became known as the *evolutionary species concept*. Dobzhansky and Mayr both then published influential books that helped form the modern evolutionary synthesis of genetics and Darwinian evolution.

The Problem of Morphology

Mayr held, and many systematists still believe this, that there was a traditional *morphological species concept* used by museum taxonomists and the like, which used superficial and anatomical traits to distinguish species. In fact, no matter what species definition they use, all taxonomists use characters, as they are called, to identify species. Morphology was never abandoned.

Mayr's biological species concept (so-called because he contrasted it with the formal morphological species concept that he had defined into existence) became, within a short time, the standard view of *species*, taught to primary, secondary and tertiary students the world over, despite the objections of many active scientists.

The morphological conception was investigated by several historians of biology and found to be roundly mischaracterised (especially by Paul Farber, a historian of ornithology). For a start, it supposedly led Linnaeus to think that species had essences, and essences could not change, so Darwinian evolution could not occur if it were correct. Moreover, any talk of types was interpreted to mean essences, and so a 'typological' account of species was anti-Darwinian; indeed creationist! Nothing worse could be said of a biologist after the Second World War than to call them a creationist. [Not coincidentally, about this time modern creationism arose among American fundamentalist Christians.]

In fact, essences were never ascribed to species in this way in history. The confusion arose because Mayr, Simpson, and others, including philosopher of biology David Hull, conflated two distinct intellectual traditions. One was the logical tradition, in which 'species' meant something like our modern 'subset', and 'genus' meant something like our modern 'proper set'. In short, a genus was a concept (in Aristotle's terminology, a 'predicate') of a general type, and a species was a concept of a special type. The other tradition was botanical and zoological classification, which had a history of over 400 years by the time the Essentialist Story arose. Moreover, Linnaeus' single-character system meant that he had, for each rank in his scheme, a *character essentialis*, which was taken to mean the essence of the species or other taxa. But in fact it was an essential part of the definition of the *name* of the taxon. However, contrary to the story, organisms could lack the *character essentialis* so long as they were born of parents of the species. Types weren't necessarily defined in terms of essences, either. They could be, so to speak, 'typical', with variation within the group (this notion was promoted by a French entomologist, Pierre André Latrielle around Linnaeus' time).

This first explicit definition by Ray did not cause any angst among zoologists or botanists. Darwin, for example, did not have a *species problem*, but a *species question*, which had been around for some decades before he published the *Origin of Species*. And that question was, 'Where did new species come from, and how?' His answer was, of course, variation followed by natural selection leading to novel traits and relative permanence. However, this was not immediately accepted by naturalists. A French anthropologist/architect named Pierre Trémaux argued in 1865 that species were the effects of differing habitats, and that

> ... if crossing within the species is impeded by whatever cause, the favoured variety would necessarily become a species, and continue to change up to the point where crossing with the mother species would no longer produce interfertility. (*Origin*)

This view was repeated a few years later by Moritz Wagner, a geographer, journalist, naturalist and beetle collector then in Munich, who argued that geographical isolation was required for new species to establish themselves so that they did not interbreed when back in contact. Unlike Trémaux, Wagner was more of a Lamarckian than a Darwinian, and he was taken to task by August Weismann, the famous zoologist at Freiburg who held that selection, both natural and sexual, was the necessary component of all speciation. Nevertheless, the Trémaux/Wagner theory of speciation by geographical isolation was adopted in the early twentieth century by David Starr Jordan, an ichthyologist (and active eugenicist) at Indiana, who influenced many American biologists. By the middle of the twentieth century, the geographical theory, called now the *allopatric speciation* theory, was the standard view, and Darwin's own view of local adaptation, now called the *sympatric speciation* theory, was largely abandoned by these Darwinians.

In a period called by historian Peter Bowler 'the eclipse of Darwinism', roughly from Darwin's death in 1882 to about 1930, Darwin's views on species receded into the background in favour of Lamarckian ideas. Indeed, Darwin's views were characterised as species scepticism, leading to a false belief that 'Darwin did not explain the origin of species in *The Origin of Species*'. These 'neo-Lamarckians', as they called themselves, accepted the inheritance of acquired characters (traits formed by the parents could be passed on to their progeny), and that species changed as a whole, which could only be caused by an internal force of some kind, usually referred to as an entelechy or inner goal. Many famous and influential scientists, such as Edward Drinker Cope, the American palaeontologist, and Alphaeus Packard, also American and a specialist in cave biology, were neo-Lamarckians, and their views of species tended to see them as lineages that changed over time in response to the environment, in parallel to the ways in which an individual organism developed, matured, and then aged. One of Cope's neo-Lamarckian students, Henry Fairfield Osborn, an American palaeontologist and another eugenicist (the era was littered with them, particularly in the United States) wrote a book *From the Greeks to Darwin*, in which he tried to establish the lack of originality of Darwin and the genius of Lamarck.

Modern Issues

Since the Synthesis between about 1940 and 1960, there have been a few developments that have added to the complexity of this problem. First of all, science moved somewhat from Mendelian genes to molecular genes. DNA, and the variety of RNA products, along with duplications, deletions, and the whole process of molecular change (for more detail, see *Understanding Genes* in this series, by Kostas Kampourakis) and an increasing knowledge of what happens between genetic sequences being expressed and cells and organisms developing have made the simple version of the 'genes' of Mendelian genetics untenable, and possibly incoherent. This has led to some researchers looking for 'speciation genes' in the sense of some genetic sequence that causes isolation in each case of a new species budding off.

Moreover, development – the sequence of modulation of the growth and maturing of organisms – was skipped over in early evolutionary thought but has come roaring back as a key aspect of how species are made, under the tag *evo-devo* (see *Understanding Evo-Devo* by Wallace Arthur, *Understanding*

Development by Alessandro Minelli, and *Understanding Evolution* by Kostas Kampourakis in this series for more information). 'Developmental systems' are a package of mechanisms which include genes but are not restricted to them, and they are also inherited and are causes of species either staying as they are or evolving into new species. Add to this an increased emphasis on environmental challenges (*eco-evo-devo*), such as fluctuating environments that could not be anticipated by fixed gene-products, or climate change, or niche construction and inheritance (see Chapter 3), and it gets much more complex. To summarise, there are now, since the end of the Second World War, between 28 (my estimate) and 32 (Frank Zachos' estimate) active species definitions, and they place different emphases on causal and cognitive features or needs. And each year there are more, as different fields of biology find need for refinements. Definitions for bacteria, viruses, hybrids, commensuals, parasite infection isolations . . . all these and much more are in the process of being developed. The nature of species is far from settled even now.

5 Philosophy and Species

If there is an issue in a science, philosophers will attend to it. This is not new, either. Since the rise of modern science in the seventeenth century, many if not most of the problems that philosophers have addressed or formulated have arisen out of science one way or another. Books on 'the philosophy of botany' or 'the philosophy of natural history' were published from the late eighteenth century onwards, although 'philosophy' meant knowledge in those days, and included scientific thinking. Nevertheless, science has always been a productive source of new problems for philosophy to chew on.

The 'species problem' has likewise been a fertile hunting ground for philosophical debate ever since geneticist Theodosius Dobzhansky raised the topic in 1935 in the journal *Philosophy of Science* (scientists used to be happy to do professional philosophy). The species problem itself arose 30 years earlier in the collision between genetics and evolutionary biology just after the turn of the twentieth century.

Natural Kinds

As I said in Chapter 1, in philosophy there are basically three questions that define problems in any area or field: what is there? How do we know it? And what is the value of it? These are called *ontology* (or *metaphysics*), *epistemology* and *axiology*. Axiological questions are not necessarily about monetary value, but also about moral value and aesthetic value, not to mention political value. We will look at the axiology of species in Chapter 8, but first let's look at the metaphysics of species.

Since the 1950s, philosophers have made passing comments about species as *natural kinds*. A natural kind is what philosophers call a 'universal', which Aristotle defined as a category or term ('predicate') that covers two or more individual things. A law of nature is a universal. So is a statistical summary, or a general concept in a scientific theory. Any generalisation is a universal, although in metaphysics, they are usually generalisations without exceptions (no natural kind is thought to be more or less a kind, nor any part of a kind more or less a part). Since Aristotle's logical term for a kind, *species/eidos*, is a universal, it was natural for recent philosophers of science to think living species were the best example of natural kinds outside of physics. Philosophers in the Anglo-American tradition of philosophy tended at that time to focus in heavily on language, so this meant that if *species* was a universal like that, there had to be a straightforward definition that specified exactly what it was to be a species (and more importantly, to say of an organism that it was a *member* of a species).

Now the idea of 'natural kinds' itself was invented, around the middle of the nineteenth century, by mathematicians and philosophers like Leonhard Euler (who invented Venn diagrams), John Venn (who didn't, but invented shading the intersections) and John Stuart Mill, who set up many of modern philosophy's debates in his 1843 book *A System of Logic*. But later philosophers of science overlooked Mill's warning that

> . . . Genus and Species, are . . . used by naturalists [biologists] in a technical acceptation not precisely agreeing with their philosophical meaning. . . (*System of Logic* Bk I ch. vii §3)

This is a fancy way of saying that what these terms mean in logic is not what terms mean in biology. And 'natural kind' is very largely a logical or metaphysical notion, not a biological one.

Still, there was a philosophical issue here, despite the lack of a historical basis in biology. What were species metaphysically? Drawing on the supposed natural kindness of species, zoologist (and later excellent historian of biology) Michael Ghiselin proposed that species are not kinds, but individuals. This needs some explanation. An individual, in metaphysics, is something that exists with a circumscribed time and place and has something that keeps it together. It is literally in-divisible without it ceasing to be what it is. So, I am an individual, because if you divide me (in an inappropriate manner! Haircuts are okay, and maybe beard trimming) I cease to be me.

A kind, on the other hand, can exist anywhere and anywhen. The Jurassic Park *Tyrannosaurus rex* (JP *T. rex*) is not part of the same species as the ones that lived in the Cretaceous (not the Jurassic!), unless the *T. rex* species is a *kind*, not an individual. Ghiselin's individuality thesis means that the JP *T. rex* is not the same species as the original *T. rex*, as the JP *T. rex* individual is not a historical part of that original species. Other philosophers, such as Michael Ruse, stuck with the natural kinds thesis and argued that a JP *T. rex* would still be a member of its species, and some do even now, although there is a rough consensus in favour of the individuality thesis.

The kinds thesis carried with it another idea: that kinds have *essences*. *Essentialism*, as this is called, has two main claims – one, that every kind has a set of *uniquely defining features* (this can mean defining the kind verbally, or it can mean causing the kind to be what it is). On this view there are physical properties that, to use the standard formulation in philosophy, *all and only* members of the kind have. They are, in short, exclusive sets of features to that kind and no other kind. This doesn't mean that other kinds don't have some of those features, just that no other kind has them all.

The other claim essentialism makes is that a kind must have those features necessarily. That is, if being a swan involved a set of traits such as shared skeletal and behavioural features, then all swans must have these features, or they aren't swans. But there are some features – such as having white plumage – that are not necessary, not 'essential to being a swan'. In the tradition of philosophy since Aristotle, these are called 'accidental properties'. Not accidental as in the result of a crash involving roadside swans and a truck carrying white paint, but only accidental in that the features don't need to be there for the organism to be included in that kind. Some swans are black (and some are albino).

Now the story that essentialism was the 'default' view in biology before some turning point, usually the *Origin of Species*, while historically untrue (see Chapter 4), was embedded into philosophy of biology by a couple of papers in 1965 entitled 'The effect of essentialism on taxonomy – two thousand years of stasis' (parts I and II) by David L. Hull, a systematist-turned-philosopher.

Hull took Ghiselin's individuality thesis and ran with it, arguing that while essentialism (and therefore natural kinds) may work in physics and language, it simply didn't in biology. Hull, along with Ernst Mayr and many others, argued that an evolutionary account meant that species did not have essences, and so they were not kinds. He once said, 'There are no laws in biology, including that one'. This became another consensus, of sorts, until a 'new essentialism' was proposed in the 2000s (see below). According to *this* later view, species having essences doesn't stop species from changing from one essence to another, and so it is compatible with evolution.

What kind of 'essentialism' could do this? Under the 'old' kind, as soon as a single organism was born that changed any element of the essence of the species, even a little bit, it ceased to be a member of its species and a new one was made in a single generation. This would broadly undercut Darwinian evolution in favour of what German geneticist Richard Goldschmidt in the 1940s called 'hopeful monsters', or massive genetic revolutions, basically meaning that an entirely different essential kind could arise in a single stroke.

The new essentialists proposed several solutions. One that philosopher Denis Walsh has proposed is that the developmental features of organisms make adaptation possible, and so the essence (developmental machinery, as it were) is a precondition of evolution. Paul E. Griffiths, a philosopher of biology, also suggests that species have a shared developmental system, and that this is conserved in evolution leading to 'historical essences' which are due to common ancestry. Another philosopher, Robert A. Wilson, argues that species have a kind of borderline essence in the mechanisms that maintain their coherence, based on the work of Richard Boyd, who proposed the 'homeo-static property cluster' account of kinds (that is, the mechanisms that make a kind stay a kind). The essence here does not have necessary and sufficient properties but rather a consensus of properties, rather like a parliamentary vote. There are quite a number of others, but the most dramatic essentialism about species is due to Michael Devitt, who argues that individual organisms must share some (mostly genetic and somewhat messy) intrinsic traits that *make* them members of their kinds.

My own view, for what it is worth, is that essentialism in biology is very much less malignant a concept than the anti-essentialists like Ghiselin and Hull had taken it to be, and that what essentialism there is in the field is pretty tame compared to the bogeyman that was supposed to have come from Plato or Aristotle. For a start, Aristotle used a phrase rather than a defining term – 'the what it is to be' – and there can be no real doubt that there are whats-it-is-to-be a dog or cat or parrot. Mind, not all, or even most, species in the world are as homeostatic as that (think single-celled species) but let us leave that aside here. There are great similarities of organisms within species. On the other hand, there are an indefinitely large number of similarities between organisms of *different* species, and sometimes a species may vary more within that species than some members of it do from some members of other species. Biology is messy.

As part of an ontology of biology, philosophers and scientists (Mayr again was a big influence here) assumed or debated that species were real objects. What real objects or kinds in science are is also (inevitably!) debated by philosophers, but we can apply a 'good enough for government work' definition here and say that species are real if they are natural.

Epistemic Objects

But this can be read in a couple of ways. One is that species have whats-it-is-to-be that are *causal powers*, as Wilson said. But another is that the whats-it-is-to-be are due to our powers of observation and thinking. That is, that species are not so much real *things* as they are *epistemic objects*, which is how philosophers talk about the ideas that we use in getting to know the world. In short, we make species when we observe and think. This position is sometimes called *conventionalism*, for those who find -isms useful, or *conceptualism*. And these views are commonly seen to require, or accompany, the notion of the *social construction of* [scientific] *ideas*.

Conventional ideas are literally ideas that are convenient to use. So, in a science, an idea can be convenient and useful without it being a fiction. But the notion that scientific concepts are social constructions is generally seen as a Very Bad Idea – it makes science just another philosophical, social or political construction, or even a religion (a claim often made by creationists and other science-dissenters). The counter to this is that a science tends to abandon useless concepts that interfere with it doing the job. Of course, scientists are social organisms – science is not done by machines in a context-free epistemic space. It follows that of course scientific ideas are, to some degree and in some respects, socially constructed. Maybe there are aliens somewhere that do not live in social contexts (at any point) that make science in such spaces. Humans, however, do live and work in social contexts. But the fact that humans are social doesn't make their science worse at finding things out; in fact, it may be a necessary part of doing human science. If species are socially constructed along with the definitions we use to do so, they may still aid us. I do not think species concepts are universally useful, but neither are they universally misleading or fictional.

Some philosophers (starting with Ingo Brigandt at Alberta) think that *species* is an 'investigative kind'. When scientists begin the study of a new area of biology, they need to have some prior 'organising principles' to begin. Brigandt suggested that the discipline's prior theoretical ideas of what makes species, such as reproductive isolation or genetic coherence, enable a researcher to begin organising the data in their new field, so that they could continue to investigate and explain what they saw. In short, *species* is a helpful tool for doing science in a complex arena.

This sort of conventionalism doesn't involve the whole 'post-modern' notion that truth is entirely relative to a culture or community (mind you, neither do most post-modernist views.) Here, the phenomena that scientists observe are not imagined or arbitrary but are actual patterns in the empirical data. What makes this conventional is the *convenience* the kind offers for doing the science. What makes it conceptualism is that this is, in the end, a concept and not (on its own) a theoretical truth. It's sort of a hybrid between scientific realism and social epistemology, as it should be – science is done by humans who live and work in societies, but they aim to investigate the world.

Much of the debate in philosophy over the metaphysics of species echoes the debates philosophers have been having for about a thousand years in the West over universals, including

- Are species/universals *real* (mind-independent) or *nominal* (existing only as names or categories)?
- Do species/universals 'cut nature at its joints' as Plato had Socrates put it in the *Phaedo*?

- Are species/universals individuals or kinds?
- How do we know if something is a species/universal?
- Is this scientific (i.e., organised knowledge) or not? If not, what is it?
- Are species/universals unified or pluralities? That is, do they uniquely identify just one sort of thing?

This means that there are deep philosophical issues underlying something as mundane as species, but species are not philosophically unique. Many such concepts, terms and applications, in science or outside of it, have the same problems.

Species Philosophy

However, there are some unique issues in the philosophy *of science* about species. One is how to identify the limits of a species. Another is what theoretical role species play in various activities of science, if any. There are also sociological questions: are species defined differently by different social groups in science, ranging from national differences to subdisciplinary differences? How do practitioners of a science gain their ideas of species? And the biggest one, to my mind, is how much of the concept of species is simply folk taxonomy?

Species are taxonomic items. Hence their role in science must be centred on what they are important for in taxonomy, first of all. A taxonomy is an ordering of things, so species are the lowest shared rank of things ordered. But they aren't the lowest rank altogether. There are subspecific ranks such as *Varieties* (Linnaeus' subspecies term), *Races* (zoology), *Subspecies*, *Stirps* (botany), *Forms* and *Subforms* (botany and mycology), *Breeds* (domestic animals and plants), *Strains* (bacteria and viruses) and so on, but these are not common to all kinds of living things. *Species* is, or rather until recently was.

Taxonomy, in and of itself, is about putting names to things. It can be applied to atoms and other particles (the 'particle zoo' of physics), to stars and galaxies, to a database of customers, library books, or to living things and their parts. Taxonomy enables us to begin investigations of things, and to reference the results of those investigations. In database terms, species names are a record identifier, and what is learned about them are the various fields of the record being filled out. This is why species' names have to be unique. If they weren't, we wouldn't know which record to access when something is said of an ambiguous name.

But taxonomy, derided by the physicist Ernst Rutherford as 'stamp collecting', is not all there is to classification. There is also another aspect, known as 'systematics', a term that derives from the attempts in the eighteenth century to systematically list and describe the features of all plants, and later, all animals. Biologists use 'systematics' and 'taxonomy' interchangeably these days, but to continue the database analogy, systematics is the attempt to describe organisms and fill in the records in as much detail as possible, to arrange them by relationships, and to explain what is going on in the living world. Taxonomy, on the other hand, creates the record/library card. And this raises yet another philosophical issue: are species the units of any biological theories?

In my youth I was taught that there were two main aspects to science: observation and theory (I am old). Each had its own 'language', and if a word or phrase was not a term used for observation, then it had to be a term of theory. But scientists do much more than simply observe or theorise. They also deal with technology and instruments (often using curse words), with lab supplies of various kinds, with each other, and most of all they interact with their subjects. In the course of this, they introduce terms that are neither entirely observational nor purely theoretical. These 'terms of art', as they are called, are useful but not necessarily referring to things that are real in theory or observation. As I noted above for species, Brigandt calls these 'investigative kinds'. My label for terms of scientific practice is 'operative terms' and the conceptual basis for them 'operative concepts'. These are procedural words and ideas. They tend to embody the traditions, practices and protocols of the disciplines, and are usually taught by imitation or instruction at the undergraduate level. *Species* is just such an operative term and concept, in my opinion.

Looking at how the term is used in various disciplines of biology and associated fields, it appears to me that it is used in a very different way in each field, as well as in different schools, nations and subdisciplines. Evolutionary biologists use it as the 'unit' (or *a* unit) of evolution. Ecologists use 'species' as a surrogate for a range of ideas, from 'niche occupier' to 'trophic node' or organisms' position in the food web. Geneticists use it as a way of keeping track of genes and gene variants. Taxonomists and systematists use it to label and store information. Conservationists use it as an index of biodiversity (a view that is being seriously challenged right now; see Chapter 7) and to 'sell' conservation of ecosystems to politicians and society by focusing on a 'charismatic' species. Field biologists as well as birders (or twitchers, as they are called in the UK) use species as identification keys in field guides. Anatomists use species to organise an idealised structure for organisms. There are other fields and users of biology, but you get the idea. Each use of the term and concept *species* has a different, disciplinary, use, and few of them are theoretically based (perhaps the ecological and genetic uses only). And those that do have theory behind them could just as easily call them something else, like *niche occupier* or *trophic role player*, or *metapopulation*, or even *gene lineage* or *genotype*.

Adding this to the wildly diverse 'definitions' of the 'species concept', and it is entirely understandable that philosophers like John Dupré, Marc Ereshefsky, and biologists like Brent Mishler and Michael Donoghue, have long been arguing that we must be pluralistic about the term or even that it must be abandoned altogether. Mishler and I went so far as to suggest that the real and most informative object in systematics is the smallest clade currently recognised and described, and that there is no species level at all. We call it a SNaRC (*Smallest Named and Recorded Clade*).

Yet you will find few working biologists who wish to eliminate the term altogether. 'Species' is the unit of both Linnaean taxonomy and a more recent proposal, slowly gaining ground, for a new systematics called PhyloCode, based on phylogenetics but which retains 'species' as the terminal node of classification. There is a lot invested in the term *species* and its function in biological sciences. Apart from the effort of getting to know the species and genus names in your field, there is the now-three-century effort of putting the information down on paper. And then there are the recent comprehensive databases like Tree of Life, ITIS, or the Catalogue of Life, and all the field-specific databases now online, like Avibase for birds. To abandon *species* is to abandon a lot of work. Nevertheless, in philosophical terms, this remains a matter of convenience rather than an ontological issue.

Bond villain Auric Goldfinger is recorded as saying, 'Mr Bond, they have a saying in Chicago: "Once is happenstance. Twice is coincidence. The third time it's enemy action."' The merits of the character to one side, this is how systematic biology is done. One specimen may be just a variant. Two similar specimens may be a variety. Three similar specimens are an indication of a species. Likewise, one species is just a species, two are suggestive of a larger group, but three is a phylogenetic relationship. In short, to have a meaningful pattern, you need three or more cases or instances. All classification boils down to iterations of 'this is more closely related to that than that other thing'. In biology, this is why phylogenetic diagrams are used. They visually represent the relations between taxa, including species. We will discuss this in more detail in Chapter 9, but for now, consider that seeing organisms, specimens and gene sequences requires at least three, but preferably many, many more instances before we can start to identify patterns. Species involve human pattern recognition (I have always said that the most effective classifier system sits between the ears of a trained and expert taxonomist), but human pattern recognition is easily subverted by social, psychological and even economic forces.

The onus is on those who think species are more than human pattern recognition with all its observer biases to show that there is really something there. This is a topic that remains debated since Ray; four centuries of arguments. You would need to go a long way in the rest of science to find a similarly contended term and topic.

6 Finding Species

There are, says Professor Julia Sigwart, an American mollusc specialist (malacologist), species *makers* and species *users*. The former are the taxonomists, and they identify, name and record species in technical journals and store the *type specimens* (the original specimen that 'bears' the name) in museums and other collections. There are way too few of these. The latter – well, that includes everybody, according to Sigwart. She notes in her 2019 book *What Species Mean* (chapter 3) that looking out of her window she sees species of tree, animal, bird and other living things, and that this knowledge involves two main steps: knowing that something is different from other similar (or related) things; and giving it a unique name to communicate and identify it to other users, for the taxonomists are also users of species. Knowing and naming species are related activities, but not the same.

Sigwart makes what, as a philosopher, I think is a minor mistake, although not one that causes science that much difficulty. She doesn't see a species of hummingbird. She sees a bird, an organism, a *specimen*. A specimen is an example of a broader group, the species, but the relation between species and specimen is fraught. On the one hand, you cannot identify a species without the use of specimens, usually by looking at many of them to see a pattern, which is then called a species. On the other hand, without knowing the broader groups that specimen is a part of (Aves, or birds, the family Trochilidae, the genus *Calypte*) you cannot identify it as a specimen of an existing species (*Calypte anna*), nor as a specimen of a newly discovered species. In short, there is a reciprocal illumination from general knowledge to particular knowledge and back again.

Naming and identifying species from specimens and vice versa is very much a philosophical topic, but in the case of taxonomy it has a particular urgency. The use of species names and descriptions has a real-world impact. As Sigwart rightly notes, most species are rare, with only a few being very common, and of that rarity only a small fraction have been named and described. But if you are investigating a new compound taken from one species of plant or animal – usually in a toxin or venom – you really need to know that the second sample taken was gained from the same species, and not some similar or related one. Likewise, when you are bitten by a snake, you need to know which antivenin to administer, quickly. However, all this depends on how uniform a species is. If the species is the same but you have been bitten or poisoned by a variant form, then the user is in trouble. But we cannot wait for full and certain knowledge of each species before we can act in these cases, or indeed in any case, or we would never be able to act. Each identification of a species therefore is a bet that it shares properties well enough to move from knowledge of specimens in the past to the current one. We call this 'induction' in philosophy: projecting from a finite number of instances to a general rule. And it turns out that unlike, say, physics, where all electrons are exactly alike, biology is not that projectable. Every organism, every lineage, every parent–child relationship, every population, every environment is different, sometimes predictably enough, but sometimes radically. So, we are on a knife edge between knowing and using, dealing with inestimable variety in the living world.

Types, Terns and Terms

The word *type* is a simple one, and we use it in our daily life along with *sort*, *kind* and similar words. But when words find their way into scientific use, they are both sharpened and made more complex, and so it is with types. We speak of the *typical* member of some group and mean by it the mode or mean (which, if I recall my high school statistics, are not always the same) of a collection of like things. But in systematics 'type' is a loaded term. William Whewell, a colleague of Darwin and the person who coined, among many other terms, 'scientist', had this to say:

> The type-species of every genus, the type-genus of every family, is, then, one which possesses all the characters and properties of the genus in a marked and prominent manner. (*Philosophy*, Vol. 1, p. 477)

In historical practice, type refers not to a typical *form*, as Whewell suggested, but to either the first form encountered, or to the most common form. I say 'form' here to indicate that it may be at any level of the taxonomic hierarchy that a type is identified.

Take, for example, the *type specimen* (or *holotype*). A specimen is what one might expect: a sample or example of something. This is very much a case of the first one encountered. It is taxonomic practice to take the first individual, fossil or trace of a new species, and name the species by attaching the name to the specimen, which is most usually stored for researchers to inspect in an authoritative repository like a collection in a museum, herbarium or university. This can lead to some funny results. My favourite example is the brown fur seal *Arctocephalus pusillus*. Found mainly around the Cape in South Africa or the southern shores of Australia and New Zealand, this species was named as *Phoca pusilla* by Johann Christian Daniel Schreber in 1776 (the species epithet ending had to change because the genus name's gender changed), after the French name given by Buffon, *petite phoque*, or in English, 'little seal'. The type specimen, however, was a female pup, well out of its territory, and probably starved to death (see Figure 6.1). The actual brown fur seal male grows up to 360 kg and up to 2.27 m (females are around half the weight). Not little, nor cute. However, since the name is validly applied according to the rules of taxonomy, *pusillus* it remains.

A type specimen is often referred to as the 'name bearer', which raises all kinds of philosophical problems. Since the specimen is dead, it is no longer part of its living species, and since it may or may not have the traits that the species in general has, how is it that it refers to a species? Or better, how is it that the name it has refers to a species? The rough answer is that there will be something about it that is relevantly similar to other members of the species, and so it connects, so to speak, via the family networks of the population of the species, but it isn't hard to think of cases where that fails.

And what if the type specimen is lost, destroyed, or simply stolen (it happens)? Then taxonomists must designate a replacement type specimen (called a *neotype*) and in cases where no type specimen was collected, a later specimen is required (a *lectotype*). There are half a dozen other types of types, but you can see the problems: depending on a type specimen of any kind means that you can lose connection to the species. In cases where specimens cannot

Phoca pusilla Buff.

Figure 6.1   The first 'little' fur seal specimen.
What the type specimen led taxonomists to think the southern fur seal (now called *Arctocephalus pusillus*) looked like.

be preserved, confusion can reign. For example, many amoeba names and species are unable to be recovered from their early descriptions because they couldn't be preserved, and the techniques used in the early twentieth century to describe them were not as precise as today's.

A *type taxon* is some taxon that exemplifies a larger taxon (Whewell's example was the rose genus *Rosa*, the type for the family Rosaceae), and so if we are concerned at all about the reality of the Linnaean hierarchy, and most are, we should ask John Locke's question about whether these general terms are chosen and shaped for the purposes of shorthand communication rather than, as Plato had Socrates say in the *Phaedo*, cutting nature at its joints as a good butcher does. If genera are not real but conventional, then what exemplifies a genus is also just that, an example which is most similar to other taxa we have gathered into the genus. Genera are in fact not named for a typical species, either. While there is a type species for each genus, it anchors the genus name rather than the properties. Likewise, higher taxa in the various codes. Each taxon has a *type sub-taxon* (sometimes called a *nominal* taxon), at least in principle. The doing of taxonomy is less clean than the formal codes governing it (Box 6.1), by reason of limited resources.

Box 6.1 Taxonomic Codes

There are codes governing the naming and procedures for animals, plants, bacteria, fungi and so forth. They are also referred to as *nomenclature codes* and are usually referred to by the initials of their name. The overall number of kingdoms has gone from two with Linnaeus in 1738, to seven in modern codes: Animalia, Archaea, Bacteria, Chromista, Fungi, Plantae and Protozoa. There is also a 'kingdom' of Viruses.

Animals: International Code of Zoological Nomenclature (ICZN), last update 2012. Governs species and genera and some higher ranks up to 'family group'.

Algae, Fungi and Plants: International Code of Nomenclature for algae, fungi and plants (ICN), last update 2017. Used to be known as the ICBN (International Code for Botanical Nomenclature) before it was amended to include algae. Includes subspecific ranks as well. It is not entirely the same as the ICZN in terminology (e.g., in plants, phylum = division).

Prokaryotes: International Code of Nomenclature of Prokaryotes (ICNP or Prokaryotic Code), last update 2008. From species to Class.

Viruses: The International Code of Virus Classification and Nomenclature (ICVCN), last update 2021. Covers realm, subrealm, kingdom, subkingdom, phylum, subphylum, class, subclass, order, suborder, family, subfamily, genus, subgenus and species.

Phylogenetic taxonomy: A proposed International Code of Phylogenetic Nomenclature (PhyloCode) governs phylogenetic nomenclature. Latest update 2019, but not yet in use. No ranks except species. All names (possibly excepting species) are to be monophyletic groups, or clades, and have only uninomials, except where species names are concerned. This is greatly debated.

I mentioned terns above (for the alliteration of course) and so we can use that group as a case. Terns are seabirds closely related to gulls and skimmers. They fall into the family *Laridae*, which has around 100 species, 41 of which are terns. There are nine genera of terns, three of which have only one species (are *monotypic*). The six remaining genera are assigned a 'nominal species', a species which first bore the genus name. The common tern (*Sterna hirundo*) is so named because the word *Stearn* in Old English/ Frisian was used to refer to it, and Linnaeus Latinised it. The species name is actually the Latin for 'swallow' based on how it flies, and it flies very well, being found from Eurasia to north America and to Australia. *S. hirundo* is the type species for the genus *Sterna*. Now, the wide distribution of the common tern has led to the naming of four subspecies: the European *S. h. hirundo*, first identified by Linnaeus in the 1758 edition of the *Systema Naturae*, which is used as the anchor of modern species names; *S. h. tibetana* in the Himalayas, Tibet and west Mongolia; *S. h. minussensis* from Siberia to north Mongolia; and finally, *S. h. longipennis* which is found in northeast Siberia to northeast China. (I was pleased to learn while researching this that terns routinely migrate to Australia and New Zealand.) So, a term for a tern type is *Sterna*, but there are several others. Clearly one good tern serves another.

Species Boundaries

Since species were first named, there have been issues with borderline cases. For example, early taxonomists often failed to realise that females were the same species as the males (especially in birds and butterflies). These errors were fixed rather rapidly as knowledge increased, but at the time of the height of professionalisation of the biological disciplines, there were strong debates regarding what was and was not a new species. A botanist, Hewett Watson, in an 1843 review in *The London Journal of Botany* attacked a just-published *Manual of British Botany* for failing to realise that some species which the author, Charles C. Babington, had named (Watson called them *book species*) were local variants of known species (Watson's term was *natural species*). Watson wrote:

> True lists of species, both as regards their names and their distinctness in nature, are essential to [biogeographical] inquiries; and yet it is impossible to make them true, while describers of plants are so continually changing both names and species in their books. Fortunately, they cannot change the species in nature also, if permanently distinct species do certainly exist.

Versions of this debate are still active. Now, we have learned a lot about species in the 180 years since Watson's critique, but we still find that we are often unsure whether a species is natural or is the result of book compilers using bad descriptions. Single species are divided into several (like elephants or orangutans; three each) and multiple species are collapsed into a single species, or maybe a *species complex* (a diverse group that is on the edge of being several species). A case used by Ernst Mayr to show that species were complex was the herring gull (*Larus argentatus*), able to interbreed by degrees around the arctic regions of Eurasia and North America until it met but would not interbreed with the lesser black-backed gull (*Larus fuscus*). For decades this has been the prime example of the phenomenon of a *ring species* to show evolution in process. However, recent work shows that in fact this is not the case but rather represents several distinct species that happen to share genes from time to time. This is not to say there aren't ring species but only that even such a well-accepted case is controversial.

But are such tests as mitochondrial DNA differences even reliable? One popular technique, which we discussed in Chapter 2, is DNA barcoding. Invented by a Canadian laboratory at the University of Guelph headed by geneticist Paul Hebert and regulated by the Consortium for the Barcode of Life (also headed by Hebert), barcoding aims to use a small sequence of genes outside the nuclear genome in organelles such as mitochondria, and so on. The presumption is that mitochondria and the rest remain pretty well constant across species. So, two inferences become possible:

- If an organism has the barcode of a previously identified species, it is a member of that species.
- If an organism's barcode does not match any previous barcode of a species, then it is a member of a new species.

Both inferences are untested across the entire range of species, but they may be very reliable in a number of cases. Still, there are issues.

One issue is that we cannot rule out that a barcode may be shared *across* species. Suppose two species of animal or plant are so closely related they have the same barcode, but on reproductive, morphological and ecological grounds behave like different species, and indeed have been named as distinct species. This can either mean that barcodes are not reliable in these groups, or it can mean that, just on this small sequence, those species are to be split or lumped accordingly. In Figure 6.2 you can see how this works.

To expect that each species has a unique barcode across all its members seems to me to be broadly unrealistic, because nothing is stable in biology over evolutionary time. So, if a barcode specialist thinks that there must be a unique barcode, they in effect are inferring from the barcode to the species (sequence

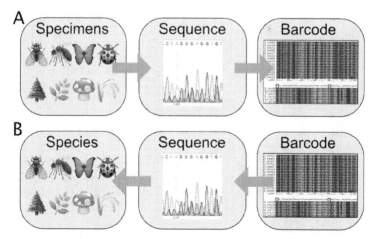

Figure 6.2 DNA barcoding.
In sequence A, the inference is from individual specimens to a barcode. In sequence B, the inference is from a barcode to something being a species. See the text for why sequence B may or may not be reliable.

B in the figure). This is a case of the tail wagging all the dogs. Barcoding is a diagnostic tool of a species (sequence A in the figure) and only gives a confident result when the barcode is *known* to be representative of a species. With the appropriate modifications to the argument, these conclusions apply to all molecular biological assays for species, including core genomes (in microbes) and 'speciation genes'.

Ever since molecular sequencing became widely feasible, the idea that genes or some other molecule could reliably identify what a species is and whether a specimen is part of a species has been bandied about. The first suggestion, that genes can tell you if you have a species, or just a variety or subspecies, is not only feasible but is being done daily, almost, or at least it seems so from the published research papers in my feed. But again, we encounter the same threshold issue of the 'Magical Molecules' section of Chapter 2: how much difference is enough? Integrative taxonomy, which includes structural, developmental and cellular information, improves things, but philosophically speaking, the problem is not solved just because one throws a slew of data at it. Whether we have 500 characters or 50,000 characters, the issues are still the same: error in measurement, failure to clearly identify homologies, subjective thresholds, and cognitive bias in the ways the algorithms that are used to analyse the data are applied.

Likewise, one of the major issues of species delimitation occurs when trying to decide if a subgroup of a named species is itself a species. Vagueness of thresholds and differences applies here more than most other applications. Often, new species are described based on molecular, usually genetic, differences (and I am not saying that this is a bad way to do it, as often we do find structure within the 'species' that we need to factor into biodiversity) but again, nothing seems to apply generally. Articles in a special issue of the journal *Fungal Diversity* in July 2021 on species concepts in fungi (the third great group of eukaryotes, along with animals and plants) noted again and again that there is no general definition of what it is to be a species, not only in fungi generally, but in subgroups of fungi like Basidiomycota, which includes many well-known mushrooms, puffballs, yeasts and so on. And these fungi reproduce sexually, to boot. So, what to do? How do we delimit species, subspecies and populations?

Part of the answer is evident in how the mycologists writing in that issue responded to this problem. Rather than relying upon theoretical concepts, instead they tend to use a variable number and type of data, ideas and practices that work well enough for each group. One proposal is to use *all* morphological, phylogenetic and molecular data in what is called the *polyphasic* approach, a protocol developed for bacteria. More generally this is referred to as *integrative taxonomy*.

As evolutionary biologist Nicholas Galtier said eloquently in 2018:

> Species boundaries ... are arbitrary while having deep scientific and societal impacts, which is not a desirable situation. In the absence of an objective solution to the species problem, different scientists analysing the same data might come to distinct conclusions and consequently issue distinct recommendations. A related issue is that, because scientists do not use a unique species definition and delineation procedure, species boundaries in different taxa have different meanings. Taxonomy is heterogeneous. ('Delineating species')

For instance, the fungal species identified are often parasitic on commercially important plants, such as rust infections of flax, linseed, coffee and so on. One of the criteria used to delimit and identify the fungus species is the host plant. But this is irrelevant to generalist fungi that can live on almost anything.

Communities

With microbial organisms and elusive macrobes, it has been very hard to isolate and describe new species. DNA barcoding is one way that has been tried to identify them, by taking a sample of, say, soil or water, and running the barcoding tests. From the number of unique markers, the number of species, known and unknown, is inferred. Many single-celled organisms cannot live alone and require some companion species. Hence, they cannot be cultured, and culturing is a traditional way to discover new microbes. Using this community-based approach (where the sample is supposed to represent the diversity of the community) has identified quite a lot of diversity in communities, and leads directly to a *metacommunity* conceptualisation of habitats, where multiple communities are seen as communicating with each other at the genetic and cellular level.

Now critics have applied the argument above that this equates species with unique barcodes rather than barcodes identifying species. And it is important not to make an epistemological virtue out of methodological limitations; if we do not have access to the organisms, we do not have access. This community barcoding is an attempt to get around that limitation, but it doesn't offer us a total resolution to microbial biodiversity. At best it indicates that we are missing many variants of the barcoding genes. Whether or not that translates to species is something only further work can establish.

There are ecosystems formed by bacteria particularly, on any surface, which quickly become multispecies communities. These are called *biofilms*, and they form on rocks, attaching themselves using specific proteins, which can also allow other distinct bacteria to attach as well, generating an *extracellular matrix*. Once this starts up, biofilms start to generate 'immune' molecules that prevent one or more species from being infected by parasites. The cells can signal each other with what is known as a *signal transduction cascade*; basically, the same mechanisms that macrobial organisms like us use to connect cells to cells. Now, biofilms are subject to many of the same problems as community barcoding: how many species are there? Techniques used to identify species sometimes involve barcodes, but also growing bacteria individually in environments like Petri dishes and incubators. One interesting issue is that at the edges of the growth media, the cells change their gene expression to form different types of cells. In natural biofilms, this also occurs; so, if species are diagnosed by the proteins they express or the morphology of the colony, this becomes another limitation on our abilities to identify species.

Conclusion

Having presented some (and only some) of the issues of identifying species, I hasten to add that all is not dark and depressing in biology. As William Yeats wrote in 'The Second Coming',

> Things fall apart; the centre cannot hold;
> Mere anarchy is loosed upon the world

Of course, he had something more metaphysical and apocalyptic in mind than the issues of taxonomy, but I wonder if he was right? Taxonomy may not be about things being delimited from each other but about centres holding together without anarchy. To use a mountain metaphor, we can easily enough see the peaks of mountains, even if we find it hard to specify the point at which one mountain ends, and another begins. Or to quote another famous author:

> Though no man can draw a stroke between the confines of day and night, yet light and darkness are upon the whole tolerably distinguishable.
> (Burke, 'Thoughts')

Species, and other taxa, which are often continuous in many ways, are on the whole tolerably distinguishable. If they weren't, we would not be having taxonomic problems, and this is the very solution several historical figures adopted, such as Lamarck or Buffon. Clusters or clumps of living things, with all their variation, gradual change, overlapping, and interbreeding, still form phenomena of things with centres, and vague borders do not change that.

7 Extinction, or How Species Are Lost

One of the things that is often said about the frankly catastrophic loss of biodiversity in the world today is that extinction is a natural process of the living world, and this is quite true. Extinction does not naturally occur at a constant rate, however. It ranges from near instantaneous (as when a 12-km-wide rock hits the planet, causing a Very Bad Day for most living things) to a slow background rate of extinction of species that have been reduced to a relic of past distributions and population numbers. So, when those who do not think we are in a catastrophic situation say, 'Extinction is natural', point out to them that the present scale of extinction is in global terms worse than a 12-km bolide, at least in geological terms, for the geological record doesn't distinguish easily between a one-day catastrophe and a four-century one. Both are 'sudden' events in Deep Time. As E. O. Wilson wrote, in his book *The Diversity of Life* (1992):

> ... life was impoverished in five major [extinction] events ... After each downturn it recovered to at least the original level of diversity. How long did it take for evolution to restore the losses after the first-order spasms? ... In general, five million years were enough only for a strong start. A complete recovery ... required tens of millions of years. ... These figures should give pause to anyone who believes that what *Homo sapiens* destroys, Nature will redeem. Maybe so, but not within any length of time that has meaning for contemporary humanity. (*Diversity of Life*, p. 31)

The Earth will survive us. The loss of species will be replenished. But it will be a Very Bad Day, for a few centuries at best and a few million years at worst, for us, and it will be of our own making, and not of the astrophysics of asteroids or comets.

What Is Biodiversity and Does It Need Species?

We have spoken of biodiversity before, and the role that species play in the legal and moral cases put for its protection and conservation, but that doesn't say what biodiversity *is* and why there is such a focus on species. And this is not as simple as the legal statutes assume it is. We discussed the example of the red wolf in the first chapter. Problematic species names and definitions aside, are species really the foundation of the measure of biodiversity?

It was entirely natural in the 1960s to equate species with the units of biodiversity. At the time, Mayr's biological species concept was predominant, and so species were thought to be unproblematically real units of nature. Counting the number of species in an ecosystem seemed to be the obvious metric, and the environmental acts passed from the 1960s onwards around the world used species as that metric. The main issue was that nobody seemed to agree on which species should be preserved. The charismatic species took priority, of course, but over time various *surrogate species concepts* (a term of conservation biology that refers to choosing one type of species to act as a surrogate for all species in a threatened environment) were developed, both to measure diversity and also to identify the species to be used.

There are basically five surrogate concepts for conservation ecology: *keystone species*, *indicator species*, *umbrella species*, *focal species* and *flagship species*. As the US Fish and Wildlife Service defines surrogate species concepts:

> species which are 'representative' of multiple species or aspects of the environment.

This is pretty broad, and these concepts play quite distinct roles in ecology and conservation.

A *keystone species* is a species that moderates the whole of the ecosystem in ways that contribute to its flourishing. It can be a species of any kind, although the case studies tend to be apex predators, like the wolves reintroduced to Yellowstone National Park, which, by reducing the number of grazers (elk, moose, deer and bison), affected the growth of forest cover, and the subsequent reduction of silt allowed the river to flow most directly, which in turn increased the beaver population, whose dams improved bird, fish and other species numbers.

A keystone species can be a herbivore, a plant, a bacterium, a fungus, a parasitic species, or even mutualistic species like pollinators (bats, bees, hummingbirds, moths). Some definitions ask of a keystone species that it play an 'essential' role in the ecosystem, which as best as I can interpret it means that it is some kind of 'critical point of failure' in the functioning of that system. If it were removed, or reintroduced after being locally extinguished, it would change the robustness of the ecosystem radically. This is therefore a *functional* notion of a species. Despite the metaphor in the title, though, if they are removed it is unlikely the entire ecosystem will collapse, although it may function quite differently than before. It is not easy to identify keystone species; often it is only after the environment degrades, or changes in a manner which we think less good, that we can say of a species that it was key. The ecologist who coined the term 'keystone species' in the late 1960s, Robert Paine, used the example of starfish, off Washington State, that kept mussels in check and thus allowed a higher degree of diversity in the area. He discovered this during an experiment (removing the starfish from an area), but nature provides such grand experiments all the time in response to overkill, whether human or natural.

I am very sceptical when so-called ecosystem engineers have designated a keystone species for conservation in the hope that the entire ecosystem will thus be protected, largely because systems usually have multiple critical components. Imagine thinking that you could keep a mechanical clock going indefinitely if you only fixed the spring every so often, despite how important the spring is for the functioning of the timepiece. Dirt in the gears, or even broken flywheels and gears, can also cause it to stop working. But this is a feature of surrogate species concepts: the hope that one species may be used to measure the health of the whole, like taking blood pressure to check all health issues. Blood pressure must be healthy for a person to be healthy, but it isn't enough, nor is good pressure an indicator of general health.

This hope is also the point of the *indicator species*. Unlike the keystone species concept, an indicator species is not supposed to cause the health of the system, but rather it is an indication of the health of the system. It's a thermometer concept. An indicator species is something that is sensitive to critical conditions like oxygen in water or temperature or loss of a food supply of general value. Some are just epiphytes – plants that live on other plants – like mosses, lichens or bromeliads. Some indicator species are apex predators such as otters or owls. Some are amphibians that are sensitive to changes in water systems, and so on. As with keystones, indicators are chosen in hope.

Keystone and indicator species do have a causal role in their ecosystems, which is what makes them useful surrogates, but several other concepts are largely conventional, or a matter of scientific limitations in resources, or purely political. That is, they are not causal concepts. They are epistemic (that is, they are due to the need to have something to hang facts on).

A popular such concept is the *umbrella species*. This is intended to be a shortcut to full analysis of the diversity of a region or a subset of the species of a region. For example, a review by New Zealand ecologists Philip Seddon and Tara Leech identified 17 different criteria for choosing an umbrella species for a region, and defined the concept as

> protection of a wide-ranging species whose 'conservation confers pro-
> tection to a large number of naturally co-occurring species' ... An
> umbrella species is in reality a population of individuals of a particular
> species whose resource requirements and habitat needs encompass those
> of co-occurring species. ('Conservation short cut')

They offered a revised list of seven criteria:

1. Natural history and ecology well known
2. Large home range size
3. High probability of population persistence
4. Co-occurrence with other species
5. Management needs benefit other species
6. Moderate sensitivity to human disturbance
7. Easily sampled or observed

It is fairly obvious that this is more than just convention. As the title of their paper put it, 'Conservation short cut, or long and winding road?' and they clearly think it is a long winding road that leads to the door of a proper notion of umbrella species. It depends a lot on what purpose people intend of the concept. If its use is to get grants from government and non-government organisations, then it can be anything the conservationist thinks will work best towards that goal. This is what a *focal species* is for. But if it is a *scientific* concept instead of a political one, then it needs to work with the science in measurable ways.

And the last of our quintet of concepts is the *flagship species*. This is a species that is great for promoting conservation in an area. It is charismatic. It is large enough for people to see it easily. It is an acceptable and indeed a likeable species. It is usually an animal (panda, koala, one of several big cats, many raptors like condors, hawks or eagles) or maybe a plant (sequoia, Wollemi pine), but it is not necessarily any of the other surrogate kinds. Flagships are usually something that fits nicely into a logo, such as the WWF's panda. Me, I like rats, bats and shrews (which account for around 70% of mammals), but I fear my personal demographic is very restricted. In any case, flagship species are primarily public relations and marketing concepts.

Measure for Measure

When science has a useful notion of the importance of something, then it must have ways to *measure* this notion, a situation in science that developed in the age of precision instruments beginning in the late eighteenth century. Nobody can deny that the notion of the diversity of life forms is a significant, even mission-critical, idea in biology and conservation, so we must have ways to measure it. And we do, in spades. In fact, the range and utility of these metrics calls into question whether biodiversity is one idea or many, and some of these are contested as being potentially useless.

The first issue is which forms of life are to be measured. As I said, species are often the units of biodiversity, but they are not the only units offered up. Others include variation within species, such as subspecies, ecological function, genetic diversity, phylogenetic diversity and ecosystem 'health'. What is more, the same mathematical formulae are often used for each of these units, so in effect there are scores of ways to assess biodiversity. Add to this human-centric interests such as Indigenous or First-Nations people and their cultural values, but also such things as 'ecosystem services' (basically an economic valuation of what ecological systems provide us), and *biodiversity* is perhaps even more complex than *species*. So, we have the question of units, the functioning of the ecosystem, the role of utilities and economics, and the rights of the species and the Indigenous peoples of the region in which the ecosystem exists. Can it get more complicated? Glad you asked. . .

There are quite a few mathematical, mostly statistical, techniques that are used by ecologists and conservationists to calculate biodiversity, in order to allocate resources to save species and ecological systems. Don't worry – I'm not going to give you the equations: after all, that is what textbooks in biostatistics and ecology are for. You do not need me to bore you; they can do it for me.

These techniques fall into the following categories:

1.   Measures of differences and similarities
2.   Measures of rarity or abundance
3.   Measures of stability and function
4.   Measures of value

and combinations of these. Differences apply to the units, so a researcher may choose to measure the rarity of some functional variant genes, or the uniqueness of species in evolutionary terms, and so on. Differences rely on setting up an abstract space (like the gene-space in Chapter 2), where some measure of similarity and distinctness can be mapped. The degree of difference is what is sometimes called *Tversky similarity* after the Israeli psychologist Amos Tversky (who died before a Nobel was awarded to his collaborator Daniel Kahneman for their shared work) who developed it. It is the number and importance of unshared items or values between two items as a fraction of their otherwise shared items or values. We can simplify this:

$$\text{Similarity}_{AB} = \text{Shared}_{AB} - \text{Unique}_A - \text{Unique}_B$$

In simple terms, the similarity between A and B is what they share, minus what they both have uniquely. Difference is the inverse of similarity. There is more to it, but that'll do us here. The thing about Tversky similarity is that it depends entirely on what Tversky called the *feature set*. This is the list of criteria, or traits, that are being tested. So, you might use the traits of, say, a domestic cow as the feature set and the traits of an eland as the target, to try to establish how closely similar they are. But which traits? This is where it gets subjective. Everything has some similarities or shared features with everything else, as the philosophers have often noted. Which are the ones that matter? Which objects should you choose as the units of similarity or difference? There are too many to select, so, metric 1 is a problem.

Well, we might then think that rarity or abundance is what counts. We want to conserve rare species rather than abundant ones, for their very rarity makes them precious. Or does it? Pollinators can be the basis of ecosystems functioning as healthy units, but pollinators are often very common. Or take fungal symbiotic relationships with plant roots, called Mycorrhizae. Many species of trees require nutrients from mycorrhizal fungi to thrive, and without them forests might not be as resilient, or even possible. Hence, rarity requires something else to establish the value of a species. In most cases, it is because rare species are regarded as valuable no matter what. In short, humans like them. This means that the rarity of a species is added to a value system, as well as facts about the interdependence of the trophic webs of the area. It is both *objective facts + subjective valuations*.

For a long time, ecologists supposed that stability of an ecosystem was the best metric of its resilience. This was the view of foundational ecologist Eugene Odum, and his brother Howard, who established the view that the best ecosystem was at equilibrium in the 1950s and introduced this assumption into American legal reasoning and policy making, thus influencing global policies. However, many recent ecologists recognise that ecosystems are constantly reacting to climatological, geographical and human-induced changes, and that the better measure of resilience is its ability to change and adapt. This implies that functional relationships between populations of species also change, and that a species may become locally extinct or reintroduced as these external conditions change.

Moreover, the massive transposition of species because of human introduction leads to the *invasive species* problem. Many such invaders were deliberately introduced to 'solve' some problem or to make migrants feel more at home. For example, the South American cane toad *Rhinella marina* (the genus name was recently changed from *Bufo*) was introduced into Hawai'i, some Caribbean islands and Australia by the sugarcane-growing British to control the cane beetle, where the toad promptly ignored the beetle in favour of many ground-dwelling birds' eggs, other amphibians, small monitor lizards and various insects, and its toxin poisons small carnivores such as the Australian marsupial the northern quoll (*Dasyurus hallucatus*). It is now so well established in Australia that it has spread from its point of introduction in northern Queensland in 1935, across the continent to the north of West Australia in 2021 (around 1,800 miles/2,800 km). From 102 breeding pairs, there are now estimated to be 200 million in less than 80 years. Incidentally, the British seem to have been very enthusiastic about spreading invasive species in their empire. Australia in particular received foxes, rabbits, deer, horses, blackberries, blackbirds and a host of other flora and fauna. We colonials do not thank them for this bounty.

As cane toads are too well established to eliminate them from their new homes, ecosystems have had to accommodate them, which raises the issue of their being functional parts of their range ecosystems. If so, do we protect them? When does an invasive species become a naturalised citizen? Consider dingoes (*Canis dingo*). They were mostly likely brought to Australia from southeast Asia or New Guinea, from Asian singing dog stock, between about 8,000 and 3,500 years ago. When they arrived, they displaced the marsupial lion (*Thylacoleo carnifex*) and the Tasmanian devil (*Sarcophilus harrisii*) from the mainland. However, dingoes are protected animals in Australia under conservation law, although, like many indigenous predators around the world, they are often shot by graziers of sheep and cattle.

In short, invasive species are basically something that humans, either Indigenous humans or colonisers, do not want in a place. They are *weeds*: a category defined by our preferences (see Table 7.1). This is demonstrated by the phrases 'good invasives' and 'bad invasives' in conservation biology. While the impacts of an invasive species may be positive or negative for the resilience of an ecosystem, that translates into good or bad because we assign it that value. As every ethics student knows, you cannot get a value statement from a statement of fact. So invasive species highlight a complexity with the *functional version of biodiversity*.

Value for All or Some?

This leaves us with measures based on value, which is, after all, what this chapter is about. Value is not an objective thing or feature of the world (unless you think everything has a purpose assigned by God, in which case this

Classification	Definition
Native	Species present by natural means
Formerly native	Species no longer present but that occurred naturally in the past
	Subdivisions:
	1. Those that could survive if reintroduced
	2. Those that could no longer survive in the present environment and climate
Locally non-native	Species introduced by humans beyond their natural geographic range
Long-established (or 'naturalised')	Species introduced by humans long ago that are now part of the food web of native species
Recently arrived	Species colonising as a result of human activities (e.g., land-use practices; human-induced climate change)
Alien	Species introduced by humans, either deliberately or accidentally

There are many definitions and categories of invasive species. This one was proposed by Charles Warren in 2007. Note that more than one category can apply in a single case, such as dingoes.

Table 7.1   Proposed classifications of invasive species

chapter doesn't matter to you as much as otherwise). It is always relative to some beneficiary. Snakes are beneficial to kookaburras, which find them a tasty snack, but they are less valuable, or even positively costly, to the smaller animals on which they prey. Most of the objections to the metrics above are based on the fact that they can result in answers that we find counterintuitive, which is to say, things we do not want to conserve, we may have to; and things we do, we may not find come out of the analysis. In short, our goals are not met all the time.

But who are the stakeholders in human evaluations? And what are they valuing species for? Often, the valuers are ecologists or conservationists, but they can run headlong into the values of traditional societies that, for example, hunt and eat sea mammals like dugongs or cetaceans, or turtles, and so on.

Which trumps which? Is the 'scientific' image of the value of things better than the 'tribal' image in any given case? Or are all such ethical standards just stand-ins for something else, like economic value?

All the attention on ecosystem services is another case of exploitative capitalism, perhaps with a somewhat gentler face than the rapacious over-exploitative kind I mentioned before. If we only value things for how they serve us, then we can devalue those that do not. And so, on it goes.

And there is a final view, which I will discuss under the rubric of Deep Ecology: the natural world is best on its own, and offers us nothing but beauty, peace and inspiration. This was what the early nineteenth-century Romantic philosophers thought, and which is still an active viewpoint today. But note, this is still wholly human-centric, and entirely subjective. As E. O. Wilson said:

> More and more leaders of science and religion now pose this question:
> Who are we to destroy or even diminish biodiversity and thus the creation?
> Look more closely at nature, they say; every species is a masterpiece,
> exquisitely adapted to the particular environment in which it has survived
> for thousands to millions of years. It is part of the world – part of Eden if you
> prefer – in which our own species arose. ('Vanishing')

However, the very term 'stakeholder' used in these debates is fraught with ways to bias the discussion. What makes someone a stakeholder? Investment, perhaps? Is there parity between stakeholders? For example, are the Indigenous people of the Amazon rainforests in Brazil of equal status with the logging and mining companies and their owners? It most certainly doesn't seem so. And even if they were, what would be the overriding values that must be addressed? Economic profit? Habitable lands? Sacred sites and organisms? It's not hard to see that in the absence of shared values, power is what determines the value of species and their habitats. And by power, I mean in the Maoist sense, from the barrel of a gun, quite literally and usually tragically.

Overkill

It is often stated that humans are the reason why large animals and birds (like the moa) went extinct. Human fossils are associated with the decline of what we call *megafauna*: mammals like mastodons, and birds like the elephant bird of Madagascar. This is called the Association Hypothesis in palaeoarcheology. However, despite it being widely accepted as fact, it is highly challenged, both in terms of the dating of the fossils, and explanatorily. A more likely explanation is a common cause – warming in the interglacial period and subsequent drying out of habitats from Eurasia to the Americas caused the decline of the megafaunal environments and at the same time enabled humans (of many species) to move into those regions.

This is not to say that some megafaunal extinctions are not due, at least in part, to human overhunting, but megafaunal hunts are rare in the palaeontological record. Humans tend to hunt smaller prey, though this may be a strategy forced upon them by the depletion of larger prey. Experimental anthropologists have shown that the tools and weapons of palaeo-Indigenous peoples are capable of killing large prey, but in general there is no real payoff. If you kill a bison or a deer, then you feed as many people in your average forager society group as you could with a mammoth. Moreover, there are real risks attached to killing large animals, risks of injury or death. Evolutionary strategies tend to avoid fitness-decreasing behaviours, and there is no reason to think that is not also true of humans and human culture.

In the end, whether or not humans have overkilled species to extinction will depend on a range of variables, such as population size, the type of ecosystems they inhabit, the range size, climate and climate changes, variability of the local weather, and pressure from other groups. In every claimed case of anthropogenic overkill, however, the evidence is at best associative and at worst counter to the hypothesis. One point worth noting, though, is that species lifetimes and fossil ranges are not identical. It takes some very special conditions to fossilise a land animal (usually in hypoxic still-water bodies like lakes, swamps and slow-flowing silty river outfalls) and so the likelihood is that if we have a late and early fossil of a species, at best that sets the minimum period the species was in existence. So, humans may have been in contact with megafauna for longer than the evidence shows, and megafauna may have persisted longer than the last known examples in the region. Correlation (in the data) is not causation.

One point about the Association Hypothesis that is not often mentioned, except by the Indigenous peoples generally blamed for the overkill, is that ascribing these extinctions solely to the first inhabitants of an area very often results in denigration of the present Indigenes and their use of the lands in which they live. In short, it can be motivated by colonialism and racism. Jared Diamond in particular has been vocal in assigning a causal role to locals for extinctions, but even the case of Easter Island and the Rapa Nui, which features so prominently in his books, is at best ambiguous and at worst already shown to be wrong. It seems to me, although on this point I am not an expert, that humans do affect their environments, constructing their niches, but then so too do other organisms, including the now extinct, and the extant megafauna at issue. What does cause overkill when it happens, however, is neither biological nor strategic. It is based, I believe, on socioeconomic factors that arrive with large populations, and therefore almost solely with agriculture. The so-called Tragedy of the Commons (over-exploitation of ecological resources for personal gain) is most likely something for which you need a highly hierarchical, that is, class-based, social structure, and therefore is unlikely to have been caused by Indigenous peoples who lacked such an exploitative society and economics. I will develop this in the next chapter.

Repairing the Damage Done?

Are there ways to repair nature? E. O. Wilson proposed setting aside one half of the planet for wilderness. This is taking the notion of wildlife parks to the limit, and it may very well solve a lot of the issue with extinction of species and even common habitats, but rare habitats in areas that humans like to live may not do so well, and likewise rare species. Balancing the space needed for human thriving and the space needed for our non-food resources is always at the heart of the problem. At present we and our agriculture sequester around 90 per cent of all living mass. This monoculture is completely unsustainable, as variety is not only the spice of life but the source of evolutionary diversity that will be needed to deal with the changes we have wrought upon our planet. So let us look at three solutions being offered. I'll ignore proposals to terraform Mars or store embryos of rare species until we fix things, because by the time that becomes feasible ecologically, I suspect we will not be in any position to make it happen.

Rewilding (or, Assisted Colonisation)

One solution is *rewilding*. This consists of the process of reconstituting degraded ecosystems and habitats with functional equivalents of the ecological actors before the region degraded. In some cases, this involves reintroduction of previously endemic species, such as wolves to Yellowstone, which has generally positive effects. This is called *assisted colonisation* and has the aim or purpose of ensuring that a single species does not go extinct. However, rewilding in general is about ecosystems, and when a major player (say a large-bodied herbivore) is lost, maybe a different one with a similar ecological role might restore the region to its primeval state. There are numerous issues with this proposal; not least is that the ecosystem in which a particular role was crucial is probably not the one that now exists in that region, and so that role may itself no longer exist or have been taken over by an introduced species and the ecosystem adapted around it instead. Second, we simply cannot accurately predict how ecosystems will behave with a major perturbation like this – maybe it will improve things (like the Yellowstone case) or maybe it will destroy things (like the cane toad case).

Third, if there are humans in that locale, immediately competing interests and power relations occur and must be resolved. A case study is the brumby, or wild horse, released by Europeans into the southern Australian alps (yes, we have mountains, and no they aren't very tall. But we can ski on them for a short period each year). Horse hooves cause bad soil erosion, as Australia has only very shallow topsoil and plants such as grasses basically hold it in place. So, there are proposals by conservation bodies to cull or even eliminate brumbies, and opposition by animal rights advocates. A similar case is the wild Arabian camel in Australia, brought in with 'Afghan' (Indian, Afghani, Egyptian and Turkish) cameleers to navigate and supply the arid centre in the 1840s. Ironically, these camels (*Camelus dromedarius*) are the last wild populations, and are regularly caught and sold back to Middle Eastern markets. A programme to reduce the population took numbers from 500,000 in 2005 to 300,000 by 2013; however, the cull has been ceased, and it seems that governments now accept camels as permanent residents in the arid deserts. The water buffalo (*Bubalus bubalis*) was also introduced about this time and is a major disrupter of monsoonal areas in the north of Australia. All were introduced (by British political and corporate interests) for economic reasons. The viewpoints of the local Aboriginal peoples who live in these regions were not sought.

A variety of rewilding is sometimes called *Pleistocene rewilding*. Tied in with the human overkill hypothesis, this involves reintroducing surrogates for large-bodied herbivores and carnivores in places where the original species died out in the Pleistocene era. That means introducing camels, horses, big cats, elephants and so forth, back into Eurasian and North American regions. The idea is that this will establish an ecological equilibrium in rewildable areas such as the Eurasian steppes, the Great Plains of America, and the boreal forests. The issue that I have with this is that (i) it is also often racist – humans were also part of that ecological era and, as we saw, it may be they moved into an area because climates were changing in the interglacial period and so these organisms were dwindling; (ii) we cannot reintroduce extinct animals, and a surrogate may not play the same role as the thing it substitutes for; and (iii) if local big cats, say, are under threat in these regions (such as the Himalayas, north America, etc.) why expect African and Asian big cats to flourish when 'reintroduced'? As for lions and rhinos in Australia, no. Just . . . no.

But some reintroductions, such as Przewalski's horse in Mongolia, work fine, in part because the ecosystem is still very much as it was when they were wild, and in part because human usage of the region is not as invasive and damaging. And that brings us to another point: the single most damaging introduced species anywhere is humanity.

Extantion (or, Resurrection)

Since *Jurassic Park* there has been a popular idea that we can 'de-extinct' organisms. Of course, not the non-avian dinosaurs, because they went extinct around 66 million years ago and DNA will survive, in optimal conditions, at most six million years. But, as the sequels showed, reviving dinosaurs, or at least the theropods (of which *T. rex* is a member), is not a smart idea. But we have also lost a lot of organisms, such as the megafauna mentioned above, within that 6-million-year window. More realistically, the half-life of DNA (the time it takes for half of a DNA strand to 'denature' or break apart) is around 520 years, so we could, with a lot of samples, maybe resurrect species that have been gone around 2,000 years or so. Asian and American megafauna, perhaps? Tiger species that are extinct? Australian thylacines (marsupial predators), which died out in 1930? And so on. This is the reason we keep hearing about mammoth replacements in Siberia and northern America. However, there are very few herbivores like mammoths, and the environment is (at present) quite different from the environments of their nearest relatives (the elephants of India and Africa). While mammoth (and mastodon) DNA is available in relatively good states, how would we create a fertilised ovum and how would we gestate that? Moreover, in viviparous (live-born) animals, the progeny gets a large amount of 'starting' gut flora, antibodies and so forth from the maternal parent.

By the way, *de-extinction* is a clumsy word. Its etymology is the negation of the putting out of a fire. But we already have a technical term for living things as opposed to no longer living things: *extant* (meaning to stand out). So, if we must make a noun from an adjective, I suggest we call it *extantion*. We cannot extant the extinct, though, I expect.

Unlike the rewilding proposal, the predators, browsers, grazers and so forth would be the ones that evolved in the habitat and ecotypes in which they would be released. We do not know how appropriately a leopard might predate upon kangaroos, but we do know that a thylacine (known as the Tasmanian wolf) will 'work' in wild Australian environments. Well, that is unless there are dingoes. But the point is clear enough. Extanting animals for the environment in which they evolved shows some promise. Moreover, I would be very pleased to see a wombat the size of a Volkswagen Beetle, the *Diprotodon optatum*, which went extinct with the drying of the Australian continent (but, alas! they died out 25,000 years ago, so no DNA is left). On the other hand, I have been chased by a very grumpy territorial male wombat, and let nobody be deceived: they are evil. So maybe it's not such a good idea to make one the size of a car. Let's have a sheep-sized echidna (*Zaglossus hacketti*) instead, if we are dreaming ambitiously!

Corridors

Most refuges for large-bodied animals are too small for them to forage for food and live in large enough groups to prevent inbreeding. It is thought that some of the more aggressive behaviours in common chimps as seen in the Gombe forest reserve by Jane Goodall in the 1970s are due to the poverty of resources and females leading to population pressure. It has also been suggested that the reason orangutans do not socialise in the wild now is because there are simply too few food resources in the forests left to them after human encroachment, and that in a richer environment they may form social groups more like other apes, as they do in rescue sanctuaries.

Population geneticists calculate that each species has a minimum bottleneck or threshold number of breeding individuals that can preserve enough genetic variance for them to survive as a species in the longer term when conditions change. This number varies according to the species' rate of reproduction and the speed of changes in its environmental challenges. This is called a *minimum viable population* (MVP). Note that this is not just the numbers of survivors of a formerly larger population, but only of those individuals that can or will be able to reproduce. In evolution, failure to reproduce or assist relatives leads to a dead end. It also makes the overall population vulnerable to sudden changes, or reductions in habitat. A case in point: most cheetahs are very closely related, some even effectively clones, because the population underwent a major bottleneck around 100,000 years ago, and possibly a second one around 10,000–12,000 years ago (including local extinction of cheetahs in North America and Europe). This bottleneck led to a dramatic drop in the genetic diversity of cheetah populations. Both bottlenecks appear to be due to the fragmentation of habitats by climate changes.

The reduction in survivability of populations with limited genetic variation is called *inbreeding depression*, because where some otherwise non-functional genes that are in a population might have been useful in somewhat different conditions, in a small population they may simply not be available. Moreover, the likelihood is that more, if not all, members of that group will be susceptible to the same diseases, such as the clone Cavendish banana and its threatened extinction by a fungal infection called Panama disease (an earlier strain of which had already destroyed a previously popular variety, the Gros Michel).

For this and other reasons, it is not a good idea to leave species in zoos or in what are called *relict populations*, remnants of a once much larger distribution. But habitats are fragmented for most wild animals as humans clear and reuse land for monocultural agriculture, industry, transport and settlement. We have done some things towards conserving species, such as maintaining 'wild' areas like national parks and forests (rarely, though, areas of economic value to humans), but even these have problems. Roads for travel and access to resource mining split habitats and kill large numbers of endangered animals. And even in countries that maintain sacred groves, the groves are usually too small for animals that cannot travel far or fly.

Some solutions are being tried. For motorways that cut through habitats, culverts to allow ground animals to cross beneath them have had a lot of success, as have wire bridges across from trees on one side to the other for arboreal animals. These allow ordinary movement within a park. But if the park lacks the resources, perhaps seasonally, what then? What if the species is migratory? Then you need to set up *wildlife corridors* of suitable habitat for migration, and not block the routes by over-clearing, over-grazing or inadvertently building border walls. These need not be continuous so long as the species can get from one to another relatively safely. Hedges are one way to achieve this. Wetlands for migrating birds are another. They aren't a complete solution, but they do contribute to one.

In countries such as Ethiopia, Eritrea, India, Nepal, Togo, Indonesia, China and Tanzania, there are local traditions of 'sacred groves' in which a considerable number of endangered species find habitat. Usually, these groves are remnants of previously widely distributed forests and ecotypes that have since been cleared for human usage, and they are simply inadequate for species to maintain viable populations. Corridors between these sacred sites, and a ban on exploitation of the sites' resources, would go a long way to retaining many endangered species. In Ethiopia, for example, the sacred sites or groves surround local churches, but encroachment on the groves by cattle farming is causing concern. They are the last examples of what are called *Afromontane* forests of the region. Sacred groves, however, are being increasingly exploited and need to be protected.

But what is it they preserve? Is it species? It appears that in fact they preserve ecotypes that were prevalent before human over-expansion. The greater diversity of species is a result, with as many as twice the number of species in the groves than in agrarian land. Whether or not it is a cause of species preservation depends on many variables, including the composition of the species involved, at all levels of the trophic network, the size and fecundity of the area preserved, and the local nature of species (widespread species are least preserved).

8 The Value of Species

What are species worth? Do they have inherent value or are they just of value to human beings? Do they have rights? Does their integrity as species have moral worth, and do we have a duty to preserve them, or to modify them? Are species of utilitarian or instrumental value? These are the questions that the third great topic of philosophy seeks to answer: *axiology* – the values of things, and the duties they impose upon us as ethical, economic and aesthetic beings.

For a long time, species have been thought to be the index marker for healthy ecosystems, for undisturbed nature and for conservation, but the reasons why have varied considerably. National Parks developed from a desire to maintain potential sources of timber, game and hunting opportunities in the United States at the end of the nineteenth and the turn of the twentieth century, as demonstrated in Teddy Roosevelt's book *The Wilderness Hunter; An Account of the Big Game of the United States and Its Chase with Horse, Hound, and Rifle*. I include the long subtitle so you may see what Teddy had in mind when he declared Yosemite Valley and Mariposa Grove a National Park in 1906. It had earlier been set aside as a park by Lincoln, but California retained control, and even though it was named a National Park in 1890, the area was not left untouched, so environmentalist John Muir convinced Roosevelt that it needed federal protection.

Later, however, this developed into what US environmentalist Aldo Leopold called the 'Land Ethic' – a species, or an organism, was held to be a good just so far as it contributed to the well-being of the land (and note: *not* human interests). He wrote, in his 1949 *Sand County Almanac*, now a classic of the environmental movement:

> The land ethic simply enlarges the boundaries of the community to include soils, waters, plants, and animals, or collectively: the land ... In short, a land ethic changes the role of *Homo sapiens* from conqueror of the land-community to plain member and citizen of it. It implies respect for his fellow-members, and also respect for the community as such.

He held that species of plants and animals, and the landscapes on which they depended, had a *right* to continued existence. The philosophical movement later called *Deep Ecology*, developed by the Norwegian philosopher Arne Naess, took this to imply that all ecosystems had a right to exist independently of the interests or rights of people. Naess believed that all species are morally equal, thus attacking the primacy of *pioneering species* (those that colonise new territories bereft of life), and by implication, later notions like *keystone species* (species that play a crucial role in the stability and well-being of an ecosystem).

When the term *biodiversity* was made popular by entomologist Edward O. Wilson in 1988 in a book with that title that he had edited (see his accessible and still generally valid 1992 book *The Diversity of Life*), the question of how to measure it became important, if only so that scientists and policy makers could tell if it was changing. Very obviously, the place to commence was the privileged status of species in the *Endangered Species Act* in the United States, and similar legislation and policies around the world. If conservation was specified by the preservation of species, then biodiversity must be assessed by how well species are preserved, and so counting species in an ecosystem or a region became crucial.

At first, the justification for conservation was to preserve potentially useful species, and these days that gets called the *ecosystem services* or in other words, what benefits humans get from a healthy ecosystem, including oxygen, CO_2 sequestration, building materials, game, fresh water and the like. This goes back at least to the Stoics (the Roman ones, anyway) but it has been tied to the western European Christian viewpoint that the world is there for the benefit of humans. Historians know this as the White Thesis, after a paper published in

1967 by a historian of technology, Lynn White Jr. The Land Ethic approach explicitly rejects this human value ethic.

The idea of ecosystem services also allows those who evaluate the state of an ecosystem to prioritise certain species or species interactions over others, such as keystone species, species that humans value, either economically or emotively, and so on. There is a well-known 'fur and feathers' bias, sometimes called the 'charismatic species effect'. These are species that are rare, endangered, beautiful, cute, impressive or dangerous. And apart from rarity, all these criteria depend upon human reactions to the species. Rarity is not enough on its own. Many rare species (slugs, for example) are not valued by humans because of their rarity. A look through the World Wildlife Fund's list of rare species (designated as 'critically endangered') suggests that if a species is not a mammal, and a large-bodied one at that, or a bird, or one of a few large turtles, it is not likely to be of concern.

So, the question arises: on what basis are species to be valued? Ronald Sandler has proposed a list of several kinds of value: *instrumental* value (what can we do with it?); *subjective* value (what do we prefer?); and *objective* value (what can we discover about its importance?), which is further broken down into three types: *natural historical value, inherent worth* and *individual organisms*.

Natural historical value relies upon something like the distinctiveness of a species. This can come from several sources: it can be the *phenotypic* distinctiveness of members of a species. A phenotype is the appearance at all scales down to the cellular structure of an organism, or basically anything that is not genetic. Or it can be *genetic* distinctiveness. This is not the genotype of the organism as such but the availability of genetic variance in a population of organisms, allowing that population to evade parasitic viruses and microbes, ecological shifts in resources, and inbreeding depression (lowered fitness caused by genetic disorders) and the like. Or it can be *evolutionary* or *taxonomic* distinctiveness. In this case, a species might be one of a few relict species in a formerly common group and have value for that reason (such as the tuatara, *Sphenodon punctatus*, the last remaining member of an order, the Rhynchocephalia, an order of reptiles – but which are not lizards – that arose in the Triassic, 250 million years ago).

Inherent worth is something that species or their members are held to have, quite separately from our interests. In a way this is like Aristotle's notion of *final* purpose, which he called a *telos*. Sandler's example is a coyote that is trapped is being prevented from being a coyote or expressing the coyote form of life. Whether or not trapping coyotes aids us, it violates the inherent value that coyotes have, their *telos*. The notion of a *telos* is far from widespread in philosophy these days, but it can be reformulated as the inherent value of the *form of life* itself. But in what does this inherent worth consist? Is it aesthetic (the form of life is beautiful or grand – every species is a masterpiece, as Wilson reported someone saying), or is it moral (we have a duty to protect the form of life), or is it functional (this species is important for the well-being of the ecosystem)? We are back to the non-objective reasons. Sandler discusses these issues in detail.

Last, we have individual organisms. It can be fairly argued that individual organisms have a good, a *telos*. They live or not. They flourish or not. They reproduce or not. And so forth. But while having a good is necessary for them to have inherent worth, and therefore to perhaps have rights, it is not enough. They must also have some feature that makes them worthy of having rights, such as sentience, or moral culpability, or even consciousness itself. These are not universal traits of all organisms, nor of all members of species that do typically have these traits. Even human beings sometimes do not have these capacities.

What all 'objective' values have, according to Sandler, is that they are not made, but are discovered. But as every introductory text on ethics points out, no amount of factual, discoverable, properties make something right or wrong, valuable or worthless. Shakespeare has Hamlet note that 'there is nothing either good or bad, but thinking makes it so'. So, if this is correct, it is our choice of these metrics that makes a species valuable. In one way this is contrary to the objectivity of 'objective' value; but in another it is not. If it is true that we humans must make a choice of metric, based on the actual situation of the species or its organisms, it may be that it would also be the case that any rational ethical being would make the same choice. While this leaves it as a choice of subjective beings, it is at least *inter*subjective, which may be as near to objectivity as we can ever approach.

There are many theories of ethics, but they are broadly divided into *consequential ethics*, like utilitarianism, in which the consequences of actions or rules make them right or wrong; *deontic ethics*, in which certain values are held to be foundational to all moral evaluations, and which involves rights and duties for moral agents (often assigned by God or Nature); and *virtue ethics*, in which natural flourishing of the moral beings sets up the virtues according to which they should live. Each type has applications in the question of the value of species. One might say that species should be permitted to live without harm (a consequentialist view), or that species just have rights (do they have duties as well?), or that species must be permitted to flourish (a virtue that is inherent for them to be what they are). And like these theories, each has its flaws. As always in philosophy, there is debate about the underlying principles: can species suffer harm? How can species have rights? Why should flourishing be important about species? And the debate goes on.

In metaethics, the study of ethical theories, there are several broader, more general, questions about the reality or objectivity of any ethical values. *Moral realism* is the view that values are discovered, not made. *Constructivism* is the view that values are made. *Error theory* is the view that we make a fundamental error of reasoning about moral realism as morals are not the kind of things that can be real or discovered. And so forth. Each of these can be applied to biodiversity issues. If you need to find out more, start with a good introduction to ethics, such as John Deigh's *An Introduction to Ethics*. Once you have done that, or if you wish to stay within the topic we are discussing, look at Sandler's 2017 *Environmental Ethics: Theory in Practice*.

The Rights of Species

One particular view about species, ethically, that is increasingly popular is that they should have rights. Some restrict this to sentient species (*animal rights*: no rights for sequoias!), some to all multicellular organisms (no rights for bacteria), but the principles are the same even if the cut-off differs. Recently, ecosystems and their parts have also been proposed for rights, under the rubric of *habitat rights*, such as a river or a forest. And to give rights is to give legal standing.

This isn't new. The first endangered species legislation in the United States was passed federally in 1966 as the *Endangered Species Preservation Act*, not long after Rachel Carson's *Silent Spring* was published. In 1973, Congress passed the *Endangered Species Act*. A worldwide convention, the *Convention on International Trade in Endangered Species of Wild Fauna and Flora*, was agreed to the same year. Similar legislation was later passed by other nations. What is new is that recently individual species are being offered rights. New Zealand's parliament granted members of the Great Ape clade limited protections roughly equivalent to the protection of a young (human) child. Other countries offer protections of various kinds against exploitation or experimentation for primates. Some European jurisdictions offer full legal personhood, which is great until one realises that some jurisdictions also offer personhood to corporate businesses.

Animal rights movements arose in the 1970s, although there are similar philosophies all the way back to the formative days of Buddhism, Jainism, and other classical early ethical movements. The topic was made popular by the Australian philosopher Peter Singer, whose books *Animal Liberation* (1975) and *The Expanding Circle* (1981) argued that restricting moral standing to just humans was wrong. Singer is a preference utilitarian (i.e., he believes we should prefer those values that tend to minimise harm and maximise happiness, rather than evaluating every single choice), but his arguments can be recast in several other metaethical ways. There have been Aristotelian (virtue ethics) animal rights movements, for example, in which the rights of species are justified by their living the lives that match their goals (*teloi*).

One strategy of Singer's that has had major impact is that he equated discrimination against species (*speciesism*) with racism, and other social discriminatory practices like the subjugation of women and the persecution of homosexuality. In this way, the properties of the species themselves are not the basis for the moral standing of species; any species has as much worth as any other. I doubt this will ever be fully applied to, for example, fungal or viral species, or disease-causing bacteria like gonorrhoea (*Neisseria gonorrhoeae*), but it is an egalitarianism of species, nonetheless.

Rights theorists ultimately ground their views upon a view of *justice*. A just action or state is one that takes into accounts our obligations to others. Under an influential view in philosophy of law, John Rawls argued the 'original position' that we should choose to make a just situation by putting up a 'veil of ignorance' about our own station in society and choosing the rights we would want in any station in our society. Can we extend this to other species? I cannot pretend I am not a member of *Homo sapiens*, so I would need to rely on something else to apply a veil of ignorance. Instead, we rely upon a veil of empathy for those species. Empathy also underlies consequentialist ethics as well, such as Singer's. This means that if you are dealing with somebody who lacks empathy for other organisms (or, indeed, other people), you cannot convince them rationally or emotionally of the justness of protecting other species. But then, one can never convince such folk of anything not in their personal interest.

One intriguing recent proposal, made by Karen Bradshaw in her book *Wildlife as Property Owners* (2020), is to assign *property rights* to other species. Under this proposal, trusts would be set up to protect the interests of a species and its environment. Again, I worry that not all species would get this kind of protection under the law, but only the fur and feathers/charismatic ones. Moreover, legal remedies are not action, but reaction. Only when a species is under threat of extinction do such remedies apply. Precautionary actions like maintaining entire habitats are much better: more effective and cheaper in the long run. Another objection I have to such legal proposals, as well-intentioned as they are, is that they prioritise formalistic and procedural actions, to the benefit of lawyers and few others, over practical and direct actions. This isn't to say that trusts for organisms, or preferably habitats and ecosystems, are a bad idea, or that laws to safeguard species, and so forth, are unnecessary. They must, though, be seen (in this author's opinion) as secondary safeguards. And that is even overlooking the possibility of the corruption of trusts by administrators and trustees to exploit the financial and other resources made available to them.

Some rights-based theorists have argued for *animal emancipation*, usually within a Marxist framework. Once more with the animal-centrism (a broader form of anthropocentrism), but at least it's egalitarian within the animal kingdom (more or less). Here also the parallel is with the oppressed (human) classes in Marx's classical theory of history and economics, such that animals are exploited classes upon whose backs (sometimes literally) capitalism runs.

How good that parallel is depends entirely upon whether or not you see animals as harmed agents whose means of production is being stolen, or animals as resources for human oppressed classes.

Capitalism and Species

Following on from historian Lynn White's 1967 argument that western European Christianity treats the living world instrumentally and without a finer appreciation for nature, a thesis that while somewhat correct is over-simplified and narrow in my view, we should look at how species are used and exploited economically.

There are several ways global capitalism has dealt with species. From its conception, capitalism has treated species as either resources – tree, fish, game species – or as incidental obstacles to doing what it does – putting roads, tramlines, gas lines and so forth through wilderness or areas under conservation management (mostly those inhabited or managed by Indigenous peoples because they have a lack of political influence, or by those who lack the resources to resist). Particularly, the oil and coal industries have caused incredible damage without much concern until it became socially and polit-ically problematic in the 1980s, and even then, these industries continued to practise indifference whenever they could.

Giving legal rights or ownership to species strikes me as a way to ensure that capitalism is able to continue exploiting through legal manoeuvres, and contractual agreements with parties other than those involved (such as trustees as surrogates for species). However, the justice-based views are not doing particularly well either, and as we noted empathy is lacking in the most egregious cases such as the destruction of the Amazon for logging and mining, or native forests and peat bogs in southeast Asia to grow palm-oil-producing plants. It is almost as if taking a self-interested approach to commerce doesn't lead to concern about the common wealth. Capitalism must be restrained in national or community interests by governments. This isn't new. Adam Smith himself allowed governments to play such roles.

One thing that can't be challenged, not realistically, is that capitalism has caused undue amounts of biodegradation, loss of biodiversity, and harm to communities that live in ecosystems that are fragile and failing. Although apologists do claim that capitalism will 'save' species like wolves, the net result of capitalist industries is over-fishing, over-logging, over-irrigating, and all the effects such diversion of natural resources (even that name is economically biased) causes. Partly this is because of the duration of the reward system in capital markets versus in ecology. Profits are near-instantaneous, whereas the ecological costs of ecosystem collapse are deferred, sometimes for centuries. However, it has now been centuries, and the bill is coming due. Fossil fuels, mining, massive agribusiness, transportation and so on are all building up costs that those who caused them either will not have to pay or are unwilling to pay. Those that carry the burden include members of those ecosystems.

Note, my target is not democratic capitalist society, which is possibly an oxymoron anyway. Communist and authoritarian political systems can be exploitatively capitalist as well. One has only to look at the record of the USSR, with the shrinking of the Aral Sea, industrial pollution, denudation of large tracts of land and so on, or China's use of coal and its exploitation of natural reserves. So long as the reward ('profit') permits the total exploitation of ecosystems, I count it as exploitative capitalism. The political arrangements under which it occurs, even if called socialistic or communist, are beside the point if it allows exploitation of this kind for profit.

This is sometimes called the Tragedy of the Commons, after an example developed by the economist Garrett Hardin. Here the individual rewards of sheep owners in competition with other sheep owners lead to an exploitative overuse of the common resources (the common lands of pre-industrial England), resulting in denudation. Hardin intended this to be an argument for private ownership not only of the animals, but of the land. He has been rightly criticised for failing to point out that common wealth is protected by common sanctions. By moving to full privatisation, the reward structure imbalance has led to the Tragedy, as individual rights are held over and above common rights, either by policy or corruption. Hence, orangutan species will become extinct as their habitat is over-exploited for profit from palm oil plantations (benefiting those in government and their backers).

This reward structure is extremely hard to shift once in place. Focusing conservation efforts and resources on individual flagship species like orangutans seems not to work, as the desire for profits from these plantations is not really influenced by their plight. The authorities who might inflict negative sanctions of weight on those who destroy the entire habitat are the same classes of people who benefit from the exploitation. There is an urgent need to penalise – that is, change the reward structure from immediate to long-term, and punishments from a nuisance to an enterprise- and personal-life-destroying sanction – and manage entire ecosystems, however defined. To achieve this would require removing the ability of law makers and enforcers to profit from the abuses. I have no suggestion for how this might be achieved politically. I think it must be part of a much broader movement.

9 Replacing 'Species'

As I have noted, terms for species are at best *polysemic* (that is, they are a single word in a language with multiple and often incompatible meanings), and at worst species is a term with no meaning of any real scientific importance. Now we will consider several replacement concepts, and the evolutionary and genetic considerations that make them seemingly viable.

In Chapter 2 we considered the extent of the different definitions as applied to a simplified version of human evolution and genetics. One of those definitions included a historical aspect – *monophyly*. Now, monophyly means either that a group is all the descendants of, and including, a single ancestral group (*historical monophyly*); or it means that an analysis of the homological traits shows that members of the group are more closely related to each other than to an outgroup (*diagnostic monophyly*). In short, a taxonomic group is based on shared identity of traits, or of lineages. The trouble with the first, historical and causal version, is that we do not have access to the past – except in the case of Richard Lenski's bacteria evolution experiment, running since 1988 at Michigan State, where he and his team have taken genetic samples from every generation since the beginning for over 70,000 generations; the bacteria have evolved (among other things) the ability to subsist on citrate, which *Escherichia coli*, a gut bacterium, cannot do in the wild. But for most evolutionary pathways, known as *lineages*, we are hypothesising about their history. While monophyly may be a historically real property, we only have access to the relationships in the data. In *integrative taxonomy*, a research programme that involves using all genetic, morphological, ecological and palaeontological data to classify, monophyly is something that explains the relations between organisms, not something we know beforehand and can use as an explanatory move.

Monophyly/single-origin diagnoses represent actual patterns in real data with a non-arbitrary set of criteria. And the smallest clade one may identify, using what resolution is in the data that we presently have, is the smallest most 'natural' taxon, in the sense that we are not concocting some kind of useful fiction (which is what scientists fear social constructivist views do). So, for a taxon to be 'real' in that sense it must be monophyletic, and as we have seen, numerous species are not. In fact, I would hazard a guess that around a third or more of recognised sexual species have genetic introgression/hybridisation, and of course no asexual species is monophyletic unless every member is a clone, directly or derived, or the species comprises a single individual.

But if we require, as the phylogenetic species definition does, that a species is monophyletic, then we have replaced 'species' with 'clade', or something similar, and the issue with clades is that they have no consistent level, no *rank*. A clade could be all eukaryotes, for example, or it could be all parrots, or a population of parrots in a single region, and so forth. This led a couple of polychaete (velvet worm) specialists, Frederik Pleijel (Göteborg University) and Greg Rouse (University of California San Diego) to propose that species should be replaced with what they termed *least inclusive taxonomic units* or LITUs. A LITU is defined as

> ... statements about the current state of knowledge (or lack thereof) without implying that they [taxa] have no internal nested structure; we simply do not know if a given LITU consists of several monophyletic groups or not. ('Least-inclusive taxonomic unit')

Now, this is a definition of naming groups – that is, it is a *nomenclature* proposal, intended to name clades that are the smallest currently identified. Brent Mishler (University of California Berkeley) and I have suggested a causal (historical) version which Brent cleverly called SNaRCs (Smallest Named and Registered Clades) to replace species as measures of biodiversity and of taxonomy. Here, the view is that any lineages that have clades at the 'least' level are the groups that we need to be dealing with. Should we find that bigger clades have smaller clades within them, then those smaller ones become the SNaRCs. The major difference between LITUs and SNaRCs is that the latter has to do with historical and causal lineages and is tied to some extent by a registry of clade names. This would require a change to nomenclatural procedures, all of which currently depend on species as a rank, and clades have no fixed ranks, just greater or lesser scope.

Figure 9.1 shows how this works, but as it stands this is not an evolutionary tree. A cladogram is a representation of *relationships* as they show up in the data. It represents the structure of the data, but any structure may be achieved in numerous ways. A lineage (a parent–child sequence of organisms, populations or other living objects) may include many unknown entities (and all the lines here represent lineages). A clade like B or C may have numerous clades within it. All we can say based on our data for now is that this structure represents the topology of the data. That is, its shape is the ways in which organisms connect. But the evolutionary history may or may not have proceeded as the cladogram suggests. For instance, A to C to G may be one unchanged lineage while B to D and E may have much finer structure in it, along with lateral genetic transfer, and so on.

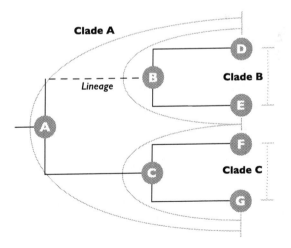

Figure 9.1   Lineages and clades.
A lineage is a sequence over time of reproductive populations of organisms. A clade is the group of the last ancestor (such as A, B or C) and all its descendants.

Now the LITU and the SNaRC proposals *are* evolutionary in the sense of there being some historical process that needs the freedom of no-rank names to deal with. Moreover, as the data comes in and our techniques for analysing them improve (like ancient DNA sequencing), there is going to be a lot of revision. And classifications like to be stable, or at least classifiers like them to be. So, what happens when species are replaced with these phylogenetic unranked ideas? Well, we have a long-term experiment for taxonomic instability: the Linnaean system.

The Linnaean system is not really a single system at all, but different systematic applications to different organisms, and each of them has its own rules on changing names. One way is to reassign the species name (the *epithet*) from one genus to another. For example, we discussed *Aspidoscelis uniparens* before, the asexual whiptail lizard. It was moved from another genus *Cnemidiphorus* in 2007, so that the North American whiptails were distinct from the older genus in South America, since they were now considered distinct clades. This happens frequently, when a group is revised in the light of new information, or changes in taxonomic techniques (molecular biology in particular) or just bad taxonomy to begin with that needs to be fixed. In one group I once reviewed, the pinnipeds (walruses, eared seals and true seals), over 40% of species had been renamed or placed into new genera since Linnaeus first named the genus *Phoca*. This is not unusual. A study by Geoffroy and Berendsohn in 2003 suggests that 45% of all 'taxonyms' are revised. Another way names and taxa can change is by either dividing a species into two or more (which happened for orangutans and the leopard frogs) or combining them because there is insufficient reason to keep them separate.

Linnaean taxonomy, whether by design or history, gave a lot of authority to the specialists who, like Richard Owen at the Natural History Museum in London in the nineteenth century, were tasked in museums and universities with describing and naming species and genera. A convention arose that when a group had been described and published, particularly by a leading figure, it was an act of discourtesy and even hubris to redescribe them, no matter how badly they were done in the first instance. This tended to mean that when the Authority died, a group might be revised in total or in part. Incidentally, the Linnaean scheme was designed to stop the taxonomic instabilities of previous schemes, and to prevent gardeners and amateur botanists from naming any different variety a new species.

The point of all this is that taxonomic instability is already there, Linnaean or phylogenetic, evolutionary or morphological. So it isn't, I think, an objection that clade-based classifications are subject to instability. Isn't a science supposed to revise its results and hypotheses in the light of better evidence?

These are not the first attempts to supplant species in classification or in biology as a whole. Beginning in the early 1900s, with the rise of Mendelian genetics, a slew of replacement terms was proposed – *jordanons*, *linneons*, *demes* (a 'neutral' taxon concept that was later taken over by the geneticists to mean a breeding population), *operational taxonomic units*, *phenoms*, *syngens* and *metapopulations*, all of which were either retasked or dropped off the radar.

The most influential non-phylogenetic replacement for species, however, is the *operational taxonomic unit*, or OTU. In the late 1950s, as computers became available for ordinary scientific work for the first time, a field known by its proponents as *numerical taxonomy*, but which later became known as *phenetics* (from the Greek *phaneros* for 'visible'; the same root as *phenomenon*), was developed, and its stated purpose was to make classification theory-neutral and objective. Since species were described in a partly subjective fashion in authoritative institutions, species as used were clearly not the way to go for phenetics, although their OTUs often correlated with the existing species names.

But the extreme empiricism of phenetics, where all evidence was used no matter what, so long as it was independent of other evidence, led to extreme instability of OTUs as a basis for classification. When, in the light of new evidence, taxonomists change the variables in the principal axes of the graphs, they can get quite distinct classifications. This is a real issue in palaeontology, when partial specimens are suddenly updated with newer and more complete specimens. A recent proposal to name a new small tyrannosaurid foundered on the argument that the specimens were just juveniles of the larger forms like *T. rex*. Hence, a species could be in several distinct OTUs at the same time (a real no-no for taxonomy). For example, using phenetic or similarity-based measures on, say, skull shape, and then using those same techniques on genetic similarity could easily lead to putting a specimen in distinct groups if the genes do not vary much but the skull shape does. Moreover, since OTUs are not ranked, one OTU could fit into several other OTUs. A species could be partly or wholly in several OTUs as well, so an OTU is not a species, nor vice versa.

The mathematics of numerical taxonomy, however, were very well worked out, and are still used even by their competitors, the cladists. The major difference is that cladism uses variations in *homologies* (for instance, tarsal and carpal bones in vertebrates). A homolog could be presence or absence of a trait that other clades had in the groups in which the analysed specimens were placed. And unlike in phenetics, in cladistics it was not *similarity* that divided or collected specimens, but *different states* of *identical* parts, such as teeth or seed structures.

Notice what the taxonomists are dealing with: not species, but specimens. That is, individual organisms, or their remains. And this has been the practice, as we noted before, with type specimens. The species name is applied first to the type specimens, and by extension to the group or taxon. In philosophical terms, a specimen is a *particular* while the general group is the *kind*. Maybe taxonomists like Buffon, Lamarck, and those who are misleadingly called *species nominalists* since, are right. Species are generalisations from particular concrete objects and thus are abstractions.

But what *is* the abstraction? The concept of a general term, the term itself, or the thing that the term and concept refer to? In order (according to me), yes, no and sometimes. A concept of a general kind, which Aristotle called 'universals' (well, actually, he called them *katholou*, the Greek word from which we get 'catholic'), is abstract. It exists nowhere in time or space. The term, since it is spoken or written or printed or chiselled, and so on, is concrete (that is, a physical thing), but it is concrete as a sign, not as the thing it signifies (whether as sound-wave patterns, grooves in stone, ink on paper or marks on a screen). Like a name, which is relevant to this issue, it can point to a real or imaginary place or thing – think of Sherlock Holmes. The thing is either real ('actual') or not ('fictional').

The position I take is that the concept is abstract, and the term *species* applies to no general kind of thing (other than a practice of taxonomists for historical reasons). But individual, or particular, species can be very real. I have no doubt that *Felis catus*, a specimen of which sits beside me sleeping as I write, is a real taxon. However, I have doubts about many other species, largely because I rely upon the people who study them, who themselves very often have doubts.

Do we need a replacement term? Well, we use universals all the time, and we argue about what gets included or not. Wittgenstein very famously argued that these are usually 'family resemblance' terms, such as *game*. Some (for example, Massimo Pigliucci at the City University of New York, both a biologist and a philosopher) have indeed tried to apply this family resemblance, or cluster, analysis to *species*. And while the use of *species* by scientists may indeed be a cluster concept, it doesn't follow that actual species are, although a family resemblance should be expected in a family group like a sexual species. So really, if we are not constrained by convention (which science is supposed to challenge from time to time), do we even need a concept with a single definition to classify? LITUs and SNaRCs suggest we do not. Biodiversity, for example, can be done on the complexity and distinctness of phylogenies without levels, and for palaeontology, where a good many species are based on a single specimen, this is unavoidable. You can't name species if you need to have a statistically valid sample in the case of fossils. But even with extant species, we can measure biodiversity as a function of variation among specimens, without needing there to be a rank or universal sort of kind.

There is a philosophical fallacy sometimes called 'misplaced concreteness' but which I prefer to call the 'reification fallacy' (to *reify* is to make something a thing; from the Latin *res*) because ideas always sound so much more important using classical languages (and I got it from a German philosopher named Herbert Marcuse, so it is doubly authoritative). Basically, it means that just because you have a noun (a 'thing word' as my English teacher called it) doesn't mean there's a thing that goes with it. Generally, words get their use and their meaning well before there is a scientific use. We speak of kinds and special things, which doesn't require that the things are real (such as fictional detectives in a genre). Maybe species are just figments of our cultural imagination. If so, they are remarkably persistent figments. None of the replacement concepts have survived for long in practice. But just because the noun continues is no reason to think the thing it points to is a real thing.

Given the history of the term, this is understandable: we all need to speak of kinds (genera) and sorts (species), as the philosopher John Locke noted in the seventeenth century:

> The reason why I take so particular notice of this, is, that we may not be mistaken about genera and species, and their essences, as if they were things regularly and constantly made by nature, and had a real existence in things: when they appear, upon a more wary survey, to be nothing else but an artifice of the understanding, for the easier signifying such collections of ideas... (*Essay Concerning Human Understanding*)

We all need shorthand ways to refer to the multiplicity of things in the world. Scientists need these ways more carefully and directly than other humans (except, maybe, for lawyers and theologians), but this does not guarantee that the ways that have developed, and which are referred to, have, as Locke calls it, a 'real existence in things'.

10 Concluding Remarks

The title of this book is *Understanding Species*, and I have spoken at length about what we understand species to be and to mean. Now, though, I would like to ruminate for a bit on the 'understanding' part.

To understand something is not necessarily to have the One True Answer. Human knowledge, and especially its concepts, is in a state of flux at all times. Sometimes, this is because we are learning new things about what the concept refers to, as is the so-called rule in science (it sometimes isn't). At other times it is because the concept no longer means anything (like 'phlogiston' in chemistry or 'vital force' in biology). But sometimes it is because the concept has been included into the 'what everybody knows' segment of culture. John Maynard Smith, a famous and influential British evolutionary biologist, called this the Bellman's Theorem (from Lewis Carroll's *The Hunting of the Snark*): 'what I tell you three times is true'.

Some words for concepts, like *fish*, have multiple meanings. For medieval Irish monks, it included the barnacle goose (which was thought to form from barnacles now called, for obvious reasons, the gooseneck barnacle) which could therefore be eaten (by the monastery's abbott and senior prelates) on Fridays, since it was fish, not meat. For whalers, whales were fish, and a famous court case in New York decreed that they were, so they attracted a smaller tax than other animals (see Graham Burnett's book *Trying Leviathan*). For Aristotle, crocodiles and octopuses were fish, because they 'lived in water'.

Science has no such kind concept as 'fish', but it does have a specialty called ichthyology, which studies and classifies 'fishes'. Indeed, palaeontologist

Neil Shubin has a book titled *Your Inner Fish*, since on phylogenetic grounds, we *are* fishes, just as we are vertebrates, mammals, primates and apes, and if we are included in those taxa, we are included in the taxa they are included in. Actually, the relevant fishes as such are called *tetrapods*, which we are also, and nobody complains about that, except Linnaeans. Species is not the only messed up concept in science.

American philosopher and pragmatist John Dewey (also known for his library classification system and educational proposals) once wrote:

> . . .few words in our language foreshorten intellectual history as much as does the word species. (*Influence*)

We get the concept of species from our culture, via parents asking what kind of animal or bird something is, to teachers talking about dogs and wolves, through to cultural items like toys, models and depictions of dinosaurs. We have been told it more than three times. Many more. Does this mean we all have an 'intuitive' understanding of species? I would argue that it doesn't. Specialists in a particular group of organisms (like fishes) spend years studying, observing, interacting with, and dissecting their subjects. They may acquire such an understanding of species *in that group*. The rest of us (and that includes scientists who are not specialists in that group), not so much. This makes sense in a way. If we had to study everything in our cultural milieu before we could use concepts, we'd never be able to begin (or we would all be a certain kind of philosopher: 'it depends on what you mean by. . .'). Instead, we get corrected by more experienced language users and authorities as we develop our thinking and speaking. And not everybody has a slew of biologist specialists on hand to correct their misunderstandings. Well, except that there are scientific reference books, websites and public education events, but even these can get lost among the Raymond Hosers of this world.

Hoser is an Australian amateur herpetologist (amphibian and reptile specialist), but he has no herpetological qualification from a formal institution. Despite this, he knows a lot of practical details about snakes. This is not at issue. As discussed, there are Codes for the naming of taxa in various fields. Herpetology is covered by the ICZN (see Box 6.1), and like nearly all taxonomic codes, it relies upon honest science and ethics to avoid having nonsense published. This means that under the Principle of Priority (Art. 23), whoever names a taxon in a publication first is the one credited for the name, and that name stands no matter what, unless the taxon is dissolved. Hoser publishes his own 'journal', in which he is the single most prolific author, and he takes the research of others and interprets it in a way to dissolve prior names and give his own. This is called 'taxonomic vandalism' in several (actually professional) papers discussing it. Hoser is not the only or the first amateur or heretic to do this – it has been an issue in taxonomy for over a century. But under the older view of science as a 'gentleman's game', it was expected that this would not occur.

This is very like the ways in which anti-vaccination advocates and climate-change deniers muddy the waters. What these taxonomic vandals are doing is devaluing science in order to aggrandise themselves. And this sort of thing, along with science fictional mistakes in naming aliens (not to mention Warner Brothers' cartoons), leaves most people quite confused.

So, to put this in a context of cognition, ask what it is to understand *anything*. Philosophers have discussed this, and one of the usual answers is that you understand what you can explain. But as I argued in Chapter 3, species have *many* causal explanations, and in any case a specialist need not know how a species came to be to understand it. Well, to understand aspects of it anyway. One can understand a species' ecology, physiology, development, genetics, behaviour and so on, without understanding its origin and history. Just as well really, because history is often hidden from us; no matter how confident we are that our favourite narrative is true, information gets lost over time. But that is for a different book.

To understand something in science is partly to know about the prototypes of things; and partly also to know why things are the way they are (to explain them). A scientist, however, can understand things they cannot explain. (And explain things they cannot understand, too, I suppose, but that is something about complexity in general and not just in biology.) What we have been considering in this book is our understanding, not of a particular species like *Felis catus*, but of the meaning, use, importance and philosophical issues of a *concept*; the concept of *species*. Understanding cats is important too, of course, but *species* is a load-bearing structure in science, and it carries an enormous weight. What philosophers raise is not science but the issues that science relies and rests upon. Hence, I have tried to clear up misunderstandings, not by giving definitive answers, but by making clearer the problems and the questions we must ask. I hope you are more *precisely* confused than you were.

Summary of Common Misunderstandings

There is one definition of species that applies to all organisms. Despite textbooks claiming that species are reproductively isolated populations – that those populations which cannot have fertile progeny are distinct species – many species can and do interbreed in the wild and have progeny that can interbreed back into one of their parent's species, often forming new species in the process, and it is quite common even with mammals. Moreover, many organisms are asexual, so that the interfertility definition doesn't apply to them, and yet they form something very like sexual species. Overall, the lack of a universal definition of 'species' is referred to as 'the species problem'.

Species have definitive traits, or essences. In diagnosing species, taxonomists will use a key set of traits to describe them. However, they do not need to be shared by all members of the species, or even be shared by the type specimen. Species are populations of organisms and organisms typically vary widely. Except among clonal species, where each organism is a genetic and bodily copy of their parent, no species will have a shared set of traits among all members and only that species' members.

Before Darwin, people believed species were unable to change. People have always believed that organisms are capable of change. The notion of there being a rank of living things called 'species' that could not change is an invention of the seventeenth century, and as soon as it was proposed, other naturalists started to argue that species weren't static, or fixed. This occurred more than 150 years before Darwin published his theory of evolution.

Species are the sole measure of biodiversity. Species have been used to measure biodiversity in ecosystems, but the mere fact of there being species to count doesn't tell us much about how diverse, resilient and productive an ecosystem is. Moreover, there are other choices of units to measure, including genetic and evolutionary distance and differences. Rarity of species that are 'naturally' endemic (local) to a region matters more than the bare number of species, but introduced species may make endemic species unable to survive.

Species are real kinds. The idea that species (in the sense of forms or terms in logic) exist only in the mind goes right back to the Greek philosophers, and in biology, it was a common view in the eighteenth century. To make things clear: there are species in the sense of taxa (the actual things) which, if they are properly identified, *may* be said to be real. There is also *species* the category, or as I prefer, the concept and term, which may not be. This is often discussed as the reality or conceptual conventionality of the species category. Taxa identified as species can be real taxa even if the category is a conventional construct in science. *Natural* is harder. The antonym of natural is in this context *artificial* but all ideas and words used in science are artificial to some extent, as they are terms of art made for a purpose. Still, one thing terms are made for in science is to identify or refer to natural objects.

Species are all equal in protective value. If species truly were real items and the units of ecology (which they may or may not be), then perhaps some species are more equal than others. In practice, the species that get legal and political protection tend to be those that are large, furry or feathery, popular with voters and primary producers, and so on ('charismatic species'). They may not be the most important to the well-being and resilience of an ecosystem, nor a good indicator of its health. We do not really know what species are essential for conservation. Our intuitions are not reliable, and we do not have a clear metric of biodiversity, nor do we have good theories about ecosystem resilience. Moreover, all attempts at conservation must deal with those humans and their interests who live in an ecosystem, and who rely upon the resources there. Still, under the precautionary principle, we should seek to preserve all species, not just the charismatic ones.

The concept and term for species will go away. Truly, replacing the notion of species to solve the problem has been tried, repeatedly, seeking to remove the term *species* from biological taxonomy. And despite there being anything from seven basic definitions of the species concept through to hundreds, in the end the term and concept means *a kind of living thing* at its most basic, and we do still need to talk about and explain kinds, no matter what the categorical words we use.

References and Further Reading

Chapter 1

References

Cain, A. J. (1954). *Animal Species and Their Evolution*, London: Hutchinson University Library.

Mayden, R. L. (1997). A hierarchy of species concepts: the denouement in the saga of the species problem. In M. F. Claridge, H. A. Dawah, & M. R. Wilson, eds., *Species: The Units of Diversity*, London: Chapman and Hall, pp. 381–423.

Phillips, M. K., Henry, V. G. & Kelly, B. T. (2003). Restoration of the Red Wolf. In L. David Mech & Luigi Boitani, eds., *Wolves: Behavior, Ecology, and Conservation*, Chicago & London: University of Chicago Press, pp. 272–288.

Winsor, M. P. (2003). Non-essentialist methods in pre-Darwinian taxonomy. *Biology & Philosophy*, 18, 387–400.

Winsor, M. P. (2006). The creation of the essentialism story: an exercise in metahistory. *History and Philosophy of the Life Sciences*, 28, 149–174.

Zachos, F. E. (2016). *Species Concepts in Biology: Historical Development, Theoretical Foundations and Practical Relevance*, Switzerland: Springer.

Further Reading

Beeland, T. Delene (2013). *The Secret World of Red Wolves: The Fight to Save North America's Other Wolf*. University of North Carolina Press.

Mayr, E. (1942). *Systematics and the Origin of Species from the Viewpoint of a Zoologist*, New York: Columbia University Press.

Mayr, E. (1963). *Animal Species and Evolution*, Cambridge, MA: The Belknap Press of Harvard University Press.

Sigwart, J. D. (2018). *What Species Mean: A User's Guide to the Units of Biodiversity*, Boca Raton, London: CRC Press.

Wilkins, J. S. (2018). *Species: The Evolution of the Idea*, 2nd ed., Boca Raton, FL: CRC Press.

Chapter 2

References

Baker, R. J. & Bradley, R. D. (2006). Speciation in mammals and the genetic species concept. *Journal of Mammalogy*, 87, 643–662.

Conniff, R. (2011). *The Species Seekers: Heroes, Fools, and the Mad Pursuit of Life on Earth*, New York: WW Norton & Co.

Cracraft, J. (1983). Species concepts and speciation analysis. In R. F. Johnston, ed., *Current Ornithology*, Vol. 1, New York: Plenum Press, pp. 159–187.

Cronquist, A. (1978). Once again, what is a species? In L. Knutson, ed., *BioSystematics in Agriculture*, Montclair, NJ: Alleheld Osmun, pp. 3–20.

de Queiroz, K. (1998). The general lineage concept of species, species criteria, and the process of speciation. In D. J. Howard & S. H. Berlocher, eds., *Endless Forms: Species and Speciation*, New York: Oxford University Press, pp. 57–75.

Dobzhansky, T. (1935). A critique of the species concept in biology. *Philosophy of Science*, 2, 344–355.

Dobzhansky, T. (1950). Mendelian populations and their evolution. *American Naturalist*, 74, 312–321.

Joyce, J. (1916). *A Portrait of the Artist as a Young Man*, New York: Huebsch.

Mayr, E. (1942). *Systematics and the Origin of Species from the Viewpoint of a Zoologist*, New York: Columbia University Press.

Regan, C. T. (1926). Organic evolution. *Report of the British Association for the Advancement of Science*, 1925, 75–86.

Rosen, D. E. (1979). Fishes from the uplands and intermontane basins of Guatemala: revisionary studies and comparative biogeography. *Bulletin of the American Museum of Natural History*, 162, 267–376.

Van Valen, L. (1976). Ecological species, multispecies, and oaks. *Taxon*, 25, 233–239.

Wheeler, Q. D. & Platnick, N. I. (2000). The phylogenetic species concept (sensu Wheeler and Platnick). In Q. D. Wheeler & R. Meier, eds., *Species Concepts and Phylogenetic Theory: A Debate*, New York: Columbia University Press, pp. 55–69.

The *Nature* definition of *haplotype* is online: https://www.nature.com/subjects/haplotypes#research-and-revie

Further Reading

Wilkins, J. S. (2009). *Defining Species: A Sourcebook from Antiquity to Today*, New York: Peter Lang. [In this book I give the major historical passages and references for species definitions, with a commentary on each.]

Chapter 3

References

Dawkins, R. (1976). *The Selfish Gene*, New York: Oxford University Press.

Schilthuizen, M. (2001). *Frogs, Flies, and Dandelions: The Making of Species*, Oxford: Oxford University Press.

White, M. J. D. (1968). Models of speciation. *Science*, 159, 1065–1070.

Further Reading

Dobzhansky, T. (1937). *Genetics and the Origin of Species*, New York: Columbia University Press.

Dobzhansky, T. (1970). *Genetics of the Evolutionary Process*, New York: Columbia University Press.

Eldredge, N. (1989). *Macroevolutionary Dynamics: Species, Niches, and Adaptive Peaks*, New York: McGraw-Hill.

Mayr, E. (2001). *What Evolution Is*, New York: Basic Books.

White, M. J. D. (1978). *Modes of Speciation*, San Francisco, CA: WH Freeman.

Chapter 4

References

Bowler, P. J. (1983). *The Eclipse of Darwinism: Anti-Darwinian Evolution Theories in the Decades Around 1900*, Baltimore and London: John Hopkins University Press.

Lamarck's definition can be found in:

Britton, N. L. (1908). The taxonomic aspect of the species question. *American Naturalist*, 42, 225–242.

Chambers, R. [published anonymously] (1844). *The Vestiges of the Natural History of Creation*, London: Churchill.

Cuvier, G. (1812). Discours préliminaire. In *Recherches sur les ossemens fossiles de quadrupèdes*, Vol. 1, Paris: Deterville.

Dobzhansky, T. (1935). A critique of the species concept in biology. *Philosophy of Science*, 2, 344–355.

The quotation from John Ray is from:

Lazenby, E. M. (1995). The Historia Plantarum Generalis *by John Ray: Book I – A Translation and Commentary*, Newcastle UK: University of Newcastle upon Tyne.

Linne, C. von (1964). *Systema naturae, 1735, by Carolus Linnaeus; facsimile of the first edition with an introduction and a first English translation of the "Observationes" by M.S.J. Engel-Ledeboer and H. Engel*, Nieuwkoop: De Graaf.

Maupertuis, P.-L. M. de (1745). *Vénus physique*, Paris?: La Haye.

Mayr, E. (1940). Speciation phenomena in birds. *American Naturalist*, 74, 249–278.

Osborn, H. F. (1894). *From the Greeks to Darwin*, London: Macmillan.

Trémaux, P. (1865). *Origin et transformations de l'homme et des autres étres*, Paris: L. Hachette.

Zachos, F. E. (2016). *Species Concepts in Biology: Historical Development, Theoretical Foundations and Practical Relevance*, Switzerland: Springer.

The various editions of the *Origin of Species* can be found at:

http://darwin-online.org.uk/contents.html#origin

Further Reading

Arthur, W. (2021). *Understanding Evo-Devo*, Cambridge: Cambridge University Press

Kampourakis, K. (2021). *Understanding Genes*, Cambridge: Cambridge University Press.

Minelli, A. (2021). *Understanding Development*, Cambridge: Cambridge University Press.

Pavord, A. (2005). *The Naming of Names: The Search for Order in the World of Plants*, London: Bloomsbury.

Wilkins, J. S. (2018). *Species: The Evolution of the Idea*, 2nd ed. Boca Raton, FL: CRC Press.

Chapter 5

Some Taxonomic Databases

ASM Mammal Diversity Database https://www.mammaldiversity.org/

Avibase – The World Bird Database https://avibase.bsc-eoc.org/

Catalogue of Life https://www.catalogueoflife.org/

ITIS: the Integrated Taxonomic Information System https://www.itis.gov/ 2015

The Tree of Life Web Project http://tolweb.org/

References

Boyd, R. (1991). Realism, anti-foundationalism and the enthusiasm for natural kinds. *Philosophical Studies*, 61, 127–148.

Brigandt, I. (2009). Natural kinds in evolution and systematics: metaphysical and epistemological considerations. *Acta Biotheoretica*, 57, 77–97.

Dobzhansky, T. (1935). A critique of the species concept in biology. *Philosophy of Science*, 2, 344–355.

Dupré, J. (1999). On the impossibility of a monistic account of species. In R. A. Wilson, ed., *Species: New Interdisciplinary Essays*, Cambridge, MA: Bradford/MIT Press, pp. 3–22.

Ghiselin, M. T. (1974). A radical solution to the species problem. *Systematic Zoology*, 23, 536–544.

Griffiths, P. E. (1999). Squaring the circle: natural kinds with historical essences. In R. A. Wilson, ed., *Species: New Interdisciplinary Essays*, Cambridge, MA: Bradford/MIT Press, pp. 209–228.

Hull, D. L. (1965). The effect of essentialism on taxonomy – two thousand years of stasis (I and II). *British Journal for the Philosophy of Science*, 15, 314–326; 16, 1–18.

Hull, D. L. (1976). Are species really individuals? *Systematic Zoology*, 25, 174–191.

Mill, J. S. (1843). *A System of Logic, Ratiocinative and Inductive: Being a Connected View of the Principles of Evidence, and Methods of Scientific Investigation*, London: John W. Parker.

Mishler, B. D. & Wilkins, J. S. (2018). The hunting of the SNaRC: a snarky solution to the species problem. *Philosophy, Theory, and Practice in Biology*, 10, 1–18.

Walsh, D. (2006). Evolutionary essentialism. *The British Journal for the Philosophy of Science*, 57, 425–448.

Wilkins, J. S. (2022). The good species. In J. S. Wilkins, F. E. Zachos & I. Ya. Pavlinov, eds., *Species Problems and Beyond: Contemporary Issues in Philosophy and Practice*, Boca Raton, FL: CRC Press/Taylor & Francis, pp. 105–124.

Wilson, R. A. (1999). Realism, essence, and kind: Resuscitating species essentialism? In R. A. Wilson, ed., *Species: New Interdisciplinary Essays*, Cambridge, MA: Bradford/MIT Press, pp. 187–208.

Further Reading

Godfrey-Smith, P. (2014). *Philosophy of Biology. Princeton Foundations of Contemporary Philosophy*, Princeton, MA: Princeton University Press.

Richards, R. A. (2010). *The Species Problem: A Philosophical Analysis. Cambridge Studies in Philosophy and Biology*, Cambridge: Cambridge University Press.

Sterelny, K. & Griffiths, P. E. (1999). *Sex and Death: An Introduction to Philosophy of Biology*, Chicago, IL; London: University of Chicago Press. (This book, although over 20 years old, is still the best introduction to the general issues of philosophy of biology.)

Wilson, R. A. (1999). *Species: New Interdisciplinary Essays*, Cambridge, MA: MIT Press.

Chapter 6

References

Boekhout, T., Aime, M. C., Begerow, D. et al. (2021). The evolving species concepts used for yeasts: from phenotypes and genomes to speciation networks. *Fungal Diversity* 109, 27–55.

Burke, E. (1865). Thoughts on the Cause of the Present Discontents. In *The Works of the Right Honorable Edmund Burke*, Vol. 1, Boston: Little, Brown, and Co. [1770], p. 477.

Galtier, N. (2019). Delineating species in the speciation continuum: a proposal. *Evolutionary Applications*, 12, 657–663.

Schreber, Johann Christian Daniel, Goldfuss, Georg August & Wagner, Johann Andreas (1776). *Die Säugthiere in Abbildungen nach der Natur, mit Beschreibungen*, Plates 81–165, Erlangen: Expedition des Schreber'schen säugthier- und des Esper'schen Schmetterlingswerkes.

Sigwart,J. D. (2018). *What Species Mean: A User's Guide to the Units of Biodiversity*, Boca Raton, FL: CRC Press, ch. 3.

Watson, H. C. (1843). Remarks on the Distinction Between Species in Nature and in Books; preliminary to the notice of some variations and transitions in the native plants of Britain. *The London Journal of Botany*, II, 613–622.

Whewell, W. (1840). *The Philosophy of the Inductive Sciences: Founded upon their History*, Vols. 1–2, London: John W. Parker.

Whewell, W. (1858). *Novum Organon Renovatum: Being the second part of The Philosophy of the Inductive Sciences*, 3rd ed., London: J. W. Parker and Son.

Further Reading

Depending on how deep you want to go, read

Farber, P. L. (2000) *Finding Order in Nature: The Naturalist Tradition from Linnaeus to E. O. Wilson*. Johns Hopkins Introductory Studies in the History of Science, Baltimore, MD: Johns Hopkins University Press.

or more technically

Sigwart, J. D. (2018). *What Species Mean: A User's Guide to the Units of Biodiversity*, Boca Raton, FL: CRC Press.

and for a criticism of the Linnaean system

Ereshefsky, M. (2000). *The Poverty of Linnaean Hierarchy: A Philosophical Study of Biological Taxonomy*, Cambridge, UK/New York: Cambridge University Press.

Chapter 7

References

Diamond, J. M. (2005). *Collapse: How Societies Choose to Fail or Succeed*, New York: Viking.

Kahneman, D. (2011). *Thinking, Fast and Slow*, London: Allen Lane.

Paine, R. T. (1969). A note on trophic complexity and community stability. *American Naturalist*, 103, 91–93.

Seddon, P. J. & Leech, T. (2008). Conservation short cut, or long and winding road? A critique of umbrella species criteria. *Oryx*, 42, 240–245.

Tversky, A., & Gati, I. (1978). Studies of similarity. In E. Rosch & B. B. Lloyd, eds., *Cognition and Categorization*, Hillsdale, NJ: Lawrence Erlbaum Associates, pp. 79–98.

Warren, C. R. (2007). Perspectives on the "alien" versus "native" species debate: a critique of concepts, language and practice. *Progress in Human Geography*, 31, 427–446.

Wilson, E. O. (1992). *The Diversity of Life*, Cambridge, MA: Belknap Press of Harvard University Press, p. 31.

Wilson, E. O. (2000). Vanishing before our eyes. *Time*, 155, 28–31, 34.

Further Reading

Jepson, P. & Blythe, C. (2020). *Rewilding: The Radical New Science of Ecological Recovery*, London: Icon Books Limited.

Kolbert, E. (2014). *The Sixth Extinction: An Unnatural History*, London: Bloomsbury Publishing.

Maclaurin, J. & Sterelny, K. (2008). *What Is Biodiversity?* Chicago: University of Chicago Press.

Stuart, A. J. (2021). *Vanished Giants: The Lost World of the Ice Age*, Chicago: University of Chicago Press.

Woudstra, J. & Roth, C., eds. (2017). *A History of Groves*, Oxford/New York: Routledge.

Chapter 8

References

Bradshaw, K. (2020) *Wildlife as Property Owners: A New Conception of Animal Rights*, Chicago: University of Chicago Press.

Deigh, J. (2010). *An Introduction to Ethics*. Cambridge Introductions to Philosophy, Cambridge: Cambridge University Press.

Hardin, G. (1968). The tragedy of the commons. *Science*, 162, 1243–1248.

Leopold, A. (1949). *A Sand County Almanac*, New York: Oxford University Press.

Naess, A. (1973). The shallow and the deep, long-range ecology movement. A summary. *Inquiry*, 16, 95–100.

Rawls, J. (1971). *A Theory of Justice*, Cambridge, MA: The Belknap Press of Harvard University Press.

Roosevelt, T. (1910). *The Wilderness Hunter*, New York: G. P. Putnam and Sons.

Sandler, R. (2017). *Environmental Ethics: Theory in Practice*, Oxford/New York: Oxford University Press.

Singer, P. (1975). *Animal Liberation: A New Ethics for Our Treatment of Animals.* A New York Review Book, New York: Harper Collins.

   1981. *The Expanding Circle: Ethics and Sociobiology*, Oxford, UK: Clarendon Press.

White, L. (1967). The historical roots of our ecologic crisis. *Science*, 155, 1203.

Wilson, E. O. (1992). *The Diversity of Life*, Cambridge, MA: Belknap Press of Harvard University Press.

Further Reading

Attfield, R. (2018). *Environmental Ethics: A Very Short Introduction*, Oxford: Oxford University Press.

Maclaurin, J. & Sterelny, K. (2008). *What Is Biodiversity?* Chicago: University of Chicago Press.

Oakes, J. (2016). Garrett Hardin's tragic sense of life. *Science in the Public Eye*, 40, 238–247.

Sandler, R. (2012). *The Ethics of Species: An Introduction*, Cambridge; New York: Cambridge University Press.

Chapter 9

References

Geoffroy, M. & Berendsohn, W. (2003). The concept problem in taxonomy: importance, components, approaches. In *Schriftenreihe für Vegetationskunde*, Vol. 39, pp. 5–14.

Lenski, R. E. (2022). The *E. coli* long-term experimental evolution project site. http://myxo.css.msu.edu/ecoli

Locke, J. (1997). *An Essay Concerning Human Understanding*, London: Penguin Books [first published 1689].

Mishler, B. D. & Wilkins, J. S. (2018). The hunting of the SNaRC: a snarky solution to the species problem. *Philosophy, Theory, and Practice in Biology*, 10, 1–18.

Pigliucci, M. (2003). Species as family resemblance concepts: the (dis-)solution of the species problem? *Bioessays*, 25, 596–602.

Pleijel, F. & Rouse, G. W. (2000). Least-inclusive taxonomic unit: a new taxonomic concept for biology. *Proceedings of the Royal Society of London Series B: Biological Sciences*, 267, 627–630.

The source text for phenetics is:

Sokal, R. R. & Sneath, P. H. A. (1963). *Principles of Numerical Taxonomy*, San Francisco, CA: W. H. Freeman.

Further Reading

Cole, C. J., Hardy, L. M., Dessauer, H. C., Taylor, H. L. & Townsend, C. R. (2010). Laboratory hybridization among North American Whiptail Lizards, including *Aspidoscelis Inornata Arizonae* × *A. tigris marmorata* (Squamata: Teiidae), ancestors of unisexual clones in nature. *American Museum Novitates*, 3698, 1–43.

Freshefsky, M. (2000). *The Poverty of Linnaean Hierarchy: A Philosophical Study of Biological Taxonomy*, Cambridge, UK/New York: Cambridge University Press.

Hey, J. (2001). *Genes, Concepts and Species: The Evolutionary and Cognitive Causes of the Species Problem*, New York: Oxford University Press.

If you want something meatier:

Slater, M. H. (2016). *Are Species Real? An Essay on the Metaphysics of Species*, Basingstoke/New York: Palgrave Macmillan.

Chapter 10

References

Burnett, D. G. (2007). *Trying Leviathan: The Nineteenth-Century New York Court Case That Put the Whale on Trial and Challenged the Order of Nature*, Princeton, NJ: Princeton University Press.

Dewey, J. (1910). *The Influence of Darwin on Philosophy*, Bloomington: Indiana University Press.

Maynard Smith, J. (1958). *The Theory of Evolution*, Harmondsworth, UK: Penguin Books.

Shubin, N. (2008). *Your Inner Fish: A Journey into the 3.5-Billion-Year History of the Human Body*, London: Allen Lane.

Further Reading

De Regt, H. W. (2017). *Understanding Scientific Understanding*. Oxford Studies in Philosophy of Science, Oxford: Oxford University Press.

Wüster, W., Thomson, S. A., O'Shea, M. & Kaiser, H. (2021). Confronting taxonomic vandalism in biology: conscientious community self-organization can preserve nomenclatural stability. *Biological Journal of the Linnean Society*, 133, 645–670.

Figure Credits

Figure 1.1   Wikipedia, Creative Commons Attribution-Share Alike 2.0. Photographer: LaggedOnUser 2016.

Figure 2.1   Redrawn by author from de Queiroz, K. (1998). The general lineage concept of species, species criteria, and the process of speciation. In D. J. Howard & S. H. Berlocher, eds., Endless Forms: Species and Speciation, New York: Oxford University Press, pp. 57–75, with permission.

Figure 3.1   Redrawn from Moritz, C. and Bi, K. (2011). Spontaneous speciation by ploidy elevation: laboratory synthesis of a new clonal vertebrate. *Proceedings of the National Academy of Sciences* 108 (24): 9733–9734.

Figure 3.4   Redrawn and simplified from Moritz, C. and Bi, K. (2011). Spontaneous speciation by ploidy elevation: laboratory synthesis of a new clonal vertebrate. *Proceedings of the National Academy of Sciences* 108 (24): 9733–9734.

Figure 4.1   Reproduced from the Rosenwald Collection at the National Gallery of Art, Washington DC, with permission.

Figure 6.1   Schreber, Johann Christian Daniel, Goldfuss, Georg August & Wagner, Johann Andreas (1774). *Die Säugthiere in Abbildungen nach der Natur, mit Beschreibungen*, Vol. Plates 81-165, Erlangen: Expedition des Schreber'schen säugthier- und des Esper'schen Schmetterlingswerkes; public domain.

Index

Page numbers for figures are in italics; for tables in bold

Printed in the United States
by Baker & Taylor Publisher Services